# THE BASICS OF CONSEQUENTIALISM

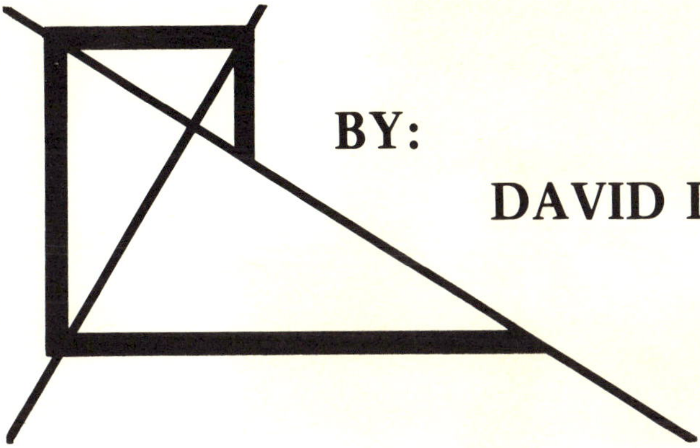

BY:

DAVID LEE CALE

With an Introduction to

## PHYSICAL PHILOSOPHY

and featuring

## THE GENESIS MODEL OF VECTON THEORY

McCLAIN PRINTING COMPANY
PARSONS, WEST VIRGINIA 26287
1980

Standard Book Number 87012-389-0
Library of Congress Card Number 80-82228
Printed in the United States of America


This book is printed for
public distribution by
Waynesburg College
Waynesburg, Pennsylvania
with funds from a
private grant



All illustrations are by the author.

*Dedicated to*
*my grandfather*
*Oliver E. Causey*

# Contents

# Foreword

Anyone interested in solving the mysteries of science and of life should enjoy reading this book. It is a philosophical and qualitative work written by a self-educated scientist, David Lee Cale. David found so many basic questions unanswered in the scientific textbooks and journals he studied that he felt a need to formulate a completely new approach to scientific thought. Further investigations will determine if his ideas are to be regarded as a step toward a better understanding of the nature of things.

David Cale's ideas are very different from views held by professionally trained scientists. He is, therefore, interested in obtaining a response from philosophers, physicists, and other professionals. Specific comments about these ideas and methods, possible weaknesses or errors, etc., will be greatly appreciated and should be sent to him. David Cale hopes that his work might stimulate further investigations.

The reader is asked to approach this work with an open mind. An application of his methodology is included, but no quantitative predictions are made. Therefore the reader should not attempt to judge David Cale's ideas by directly comparing them with other theories. Accepted theories are themselves still somewhat uncertain. This uncertainty remains, I believe, because the modern theories of physics provide few answers to basic questions; give little in the way of a common sense, qualitative picture of things; and then miraculously yield accurate quantitative predictions after pages of nearly undecipherable mathematics. Yes, the final theoretical equations agree nicely with the experimental measurements, but the correct answers were already known: the experiments came first!

Judging a new methodology is a slow and difficult process. The history of science shows that "mainstream" scientists usually reject new ideas. Eventually a few experts carefully study a new theory, and based on experimental verification, urge tentative acceptance. Other scientists gradually accept the new ideas. After a generation or two, however, the original experts are all dead and the popular theories begin to be accepted without question: the textbooks present them as "fact" rather than as "tentative theoretical guesses," which they always remain. Let us pray that there will always be people willing to risk working in the "fringes," to create their own paths, and to temper enthusiasm for the academically proclaimed science of the day by recalling the definition that "science is the belief in the ignorance of experts" (Richard Feynman, 1965 Nobel laureate). Only then can science as a whole progress.

Dr. David K. Walker, Ph.D.
*Professor of Physics*

Waynesburg College
Waynesburg, Pennsylvania
March 1980

# Preface

The philosophy of consequentialism, and vecton theory in particular, is the product of several years of work by Cale. His first effort at finding an underlying order in nature was a paper written in 1963 setting forth a planetary distance formula based on Fibonacci numbers and their attendant geometries as found in certain biologic morphologies. Because Cale's formula was more accurate than the famous Bode planetary distance formula, it served as the basis for Henry Gould's 1965 paper titled: "Cale's Rule for Planetary Distances."

In most textbooks July 16, 1945, is given as the date of the opening of the atomic age in that on the morning of that day the first atomic bomb was exploded. David Lee Cale was born that same morning in New York City, the son of a Methodist minister. David's grandfather was the developer of a large commercial cave near Uniontown, Pennsylvania, and David took over its management at the end of his freshman year in college.

The responsibilities of the business, as well as its isolation from any college, forced Cale to work toward his B.A. in political science on a part-time basis. Yet, his interest in the physical sciences remained strong.

Using Laurel Caverns, as the cave is known, as a kind of lab, Cale developed an air-conditioning system which used the cave's naturally cool air to air-condition the attraction's main service building. In 1974 he developed a heat exchanger which, using butane as the refrigerant, was efficient enough to operate a small steam engine using the potential between the constant 50°F cave air and the cold winter air of Pennsylvania. It was perhaps the first time anyone had ever derived energy by artificially merging two naturally occurring air temperature differences.

The job of managing the caverns put Cale in the mainstream of community work and of the tourist industry. By the age of twenty-one he was on the Board of Directors of the Uniontown, Pennsylvania, Chamber of Commerce; and at twenty-five he was president of the Pennsylvania Attractions and Travel Association. The numerous community offices he has held have earned him many awards and other recognitions over the years.

In 1975 David extensively toured Taiwan as a governor's representative for a Rotary Group Study Exchange. From similarities he noticed between the ancient Chinese culture and that of the ancient Mayans and modern Athapascan Indians, Cale concluded that significant migrations into North America must have taken place from Northern China, Korea, and Japan during the Chou and Han dynasties. In addition to such traditional comparisons as between temple structures, farming techniques, dress, and statuary, Cale worked out a statistical syllable comparison between proper names. That is, he takes the view that the oriental influence in American Indian cultures can be determined by the high statistical occurrence of such syllables as mei, gei, chi chen, chan, ko, shan, ni, han, and da in both ancient Asian and Indian proper names.

David Cale also developed a metrically divided circle which has several advantages over the 360° circle. He divides the circle into 100 equal parts, each called a centigree. Each centigree can then be further divided in powers of ten as the need for accuracy requires. For example, the notation $47.3640^c$ specifies a certain exact angle, while a 90° angle would be notated as $25.0^c$. His system makes directional and vector calculations much simpler; for instance:

| Classic Notation | Calean Notation |
|---|---|
| 37° 27′ 13″ | 10.403780$^c$ |
| -26° 41′ 19″ | -7.413503$^c$ |
| 10° 45′ 54″ | 2.990277$^c$ |

Cale even used the metric system when he designed and built his home. Using oak and stone right from the cavern's property, he built a kind of "tudor-gothic" house, whose interior and exterior proportions are all in meter multiples. The cale "chateau" may be the first home ever built to the metric system in the United States. There are no wires or aluminum visible on the exterior of the house, nor any material that was not available in the Fourteenth Century.

David married Lillian Gangwere of Wheeling, West Virginia, in 1977 and the two of them now reside in their unique home at Laurel Caverns along with their pet kitten, Fuzzy.

The Cale home at Laurel Caverns. Sketch by David Cale.

## Acknowledgments

A special thank you is extended to my friend, Dr. Peter Ostrander, for his thoughtful review of the first draft of this book, and to professors Wilbur Bluhm and David Walker for their very thorough reviews of my final draft. Finally to my wife Lillian I extend my most heartfelt appreciation for her patience and assistance.

David Lee Cale

# Introduction

One of the fundamental needs of thinking people is the need for a logical cosmology to explain the origin of their own existence as well as all that they perceive around themselves. For much of recorded history this need has been filled by classical philosophies or the creation narratives of popular theologies. Modern men and women have not been so fortunate. The advances in scientific knowledge achieved since the Renaissance have deprived them of the simple and orderly view of the cosmos held by many of their ancestors.

As a typical example, the belief that everything is comprised of only the elements earth, air, fire, and water was popularly held for nearly two thousand years. The longevity of this belief undoubtedly stemmed from its simplicity, symmetry, and seeming fit with the observed world. Consider, for instance, how well the concepts of wet, dry, cold, and hot fit this four-element theory.

The four-element model of the physical universe

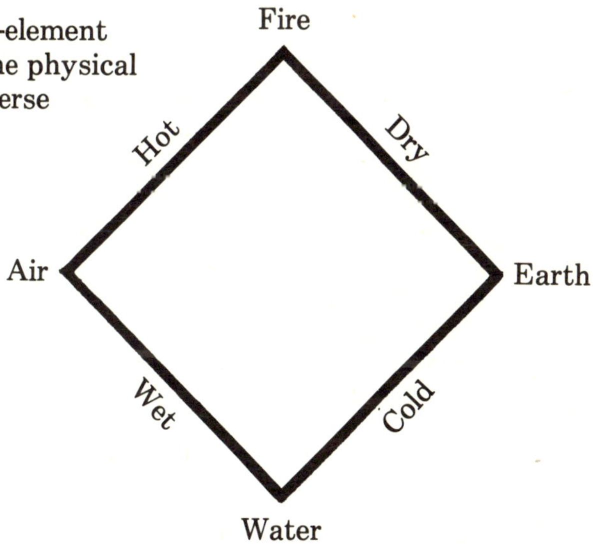

There seems to be in man an innate desire to believe that all of the diversities of the observed universe are based upon a simple set of truths. This belief is tenaciously held even today in spite of an ever proliferating list of sub-atomic structures. Christian fundamentalists work earnestly on cataclysmic creation theories to keep new scientific findings in line with Genesis, while theoretical physicists seek to simplify and unify the laws of physics. Truly, man needs a logical explanation for the complexities of his universe.

This desire probably arises from some basic need for order and predictability in life. Since success and survival often depend on how well one orders his affairs and plans ahead, the need for an orderly view of things was probably "selected in" in man's evolution along with his intelligence. Whatever its source, I know that I, and a lot of others like me, have the need.

How wonderful it would be for us to go to a minister, or priest, or philosopher, or physicist and be able to learn the nature of all that is. Better still, how wonderful it would be to have that explanation flow as an ordered sequence so that we may know the relationship between things. But we cannot do this, for no one seems to know the complete nature of reality.

Those of us not content with the currently popular views of reality are confronted with two possibilities; either we can give up on our desire for a total understanding of all that is, glibly hoping somebody will someday come up with a view of reality that makes sense, or we can confront the question ourselves, developing a model of the universe that at least makes sense to us. I chose the latter, and out of that choice grew the philosophy of consequentialism and this work.

Most of us have neither the resources nor time to attack the mysteries of the universe. We therefore usually accept the popular beliefs of our time as valid and tend to develop from them some generalized outline of reality.

For all of us these generalizations are full of unaddressed details such as the nature of primary cause or the cause of gravitation.

Any person seeking to develop for himself some ordered view of reality must first ask himself if such an attempt is possible, given the limits of individual resources and human perspective. For myself the answer to this question was yes.

Through research and reading one can glean in a matter of months the fruits of many lifetimes of work. The information of research projects that cost perhaps millions of dollars is available for the price of a few journals. Through learning one can know the thoughts of a thousand minds. My individuality does not limit the potential of my efforts. The question of human ability to eventually understand the exact nature of reality is not as easily addressed. However, for reasons that I will touch on in this book, I consider the human mind perfectly capable of knowing the nature of all physical reality.

Once an individual accepts the heady notion that he is personally capable of meaningfully addressing the mysteries of the universe, the next issue for him is the one of methodology. He must ask himself: "How can I take this tremendous and diversified array of observed facts and opinions and make any sense of it all?" By asking himself this question he takes on the role of a philosopher.

All conceptualizations about the nature of reality are arrived at through one methodology or another. The methodology might turn on some kind of spiritual revelation, scientific method, inner awareness, qualitative reasoning, moral living, or just plain common sense. The person who seeks to know the nature of reality by prayer and communion with God has his methodology just as the scientist utilizing rigorous experiments has his.

It is not likely that anyone will unify the four attractive forces in nature through meditation. On the other hand, purely quantitative reasoning may never be able to address questions relating to primary cause. The question of which method or methods are best for determining the nature of reality is basic for those individuals who believe humans can come to know what the universe is really all about.

This book is about a method which I developed to assist me in organizing and understanding the facts about the universe which I acquired through learning and experience. Whether or not you will be able to use this method for forming your own opinions about the nature of reality will depend upon the degree of satisfaction you have with existing popular methodologies and the degree to which you accept their premises.

All methodologies have a set of basic assumptions which serve to give credence to the methodology. For example, the belief that knowledge can be acquired through prayer assumes there is a God while scientific method assumes all physical laws are fixed or at least predictable within certain probable limits. Consequentialism, too, has its premises.

Probably the most basic and subtle of the premises of consequentialism is that there exists such a thing as absolute truth. The Principle of Uncertainty, a cornerstone of quantum theory, has made popular the notion that truth is always only an approximate thing. Certainly, as observers of the universe, our measurements are at best approximations . . . but does that mean the truths we are measuring are not absolute? This observer perspective is one of the topics dealt with in the first section of this book which outlines the background, premises, divisions, and methods of consequentialism.

In the second section of this book I give a sample of the first step of my methodology. The understanding of this sample requires an open mind, careful study, and a knowledge of particle physics. The sample is only offered to illustrate the novel possibilities of consequentialistic methodology and to show that the approach has potential. It is not the purpose of this book to argue the view of the universe presented by the sample reality model. Here I am concerned primarily with the method, not the product.

Now let me share with you the thoughts and opinions that led me to believe that the universe can come to be understood by the building of what I call "reality models."

# *Section I*

# The Consequentialistic Perspective

## CHAPTER ONE

# Physical Philosophy

There exists within the community of scientists the widely held belief that the arguments of science are only valid when presented in a quantitative manner. For instance, the authors of papers on any new problem or new concept in physics must back up their theses with theory comparisons and mathematical proofs. If a paper is not predicated upon its own rigorously completed experiment, then it must rest on a theory which has already gained popular acceptance.

As a way of handling direct questions bearing on physical observations, the method has been successful. This approach, as a popular form of scientific method, makes four basic assumptions about reality in most applications.

First: It assumes that there exists a reality of physical structures having fixed laws unaffected by human opinion.

Second: It assumes that mathematics is a valid tool in interpreting that reality.

Third: It assumes all factors are conserved in every interaction so that the product of any experiment is valid.

Fourth: It assumes that as a method of arriving at ultimate physical truth it is self-sufficient. That is, all questions in physics can be answered or adequately approached with, and only with, mathematical statements based upon direct or indirect observations or established theories.

The existence of this last assertion is demonstrable. At the time of this writing few, if any, physical societies will publish papers exhibiting qualitative arguments. However, it must be questioned as to whether this bias is warranted. Put another way: is the fourth assumption of this scientific method valid?

Any value judgment on its validity turns on the success physics has had in explaining the nature of reality. This is its only real test. Yet, the measure of success is itself a value judgment. Who is to say what constitutes success and what doesn't? Ultimately it is a personal evaluation which has strength only to the degree that it is shared by others. At the time of this writing, there is a modest, but growing sense among thinking people, both in and out of physics, that the science is not dealing adequately with the problems it confronts.

In addition to the fore-mentioned popular premises of scientific methods, modern physics rests upon three theories which have defied successful challenge since their promulgation in the early decades of this century. They are: the theory of special relativity, the theory of general relativity, and quantum theory.

Arguments, which are critical of the success of the science, turn on the observation that physics is still only a collection of patchwork theories. In spite of the fact that scientific methodology and the three basic theories of physics have been authoritatively dealt with for over fifty years, and though countless volumes of quantitatively precise documents have been produced by thousands of highly educated theorists spending hundreds of millions of dollars from grants and other funding sources, the laws of physics have not been unified as the premise of one physical reality suggests they should be. Further, the science has not even been successful in determining the fundamental causes of such everyday phenomena as gravitation and permanent magnetism.

There are, generally speaking, two possible explanations for this shortcoming. The first is to suggest that reality is an extremely ramified truth whose complexities have defied even the most sophisticated investigations. The second is to suggest that reality is not really all that complicated and that the failures of physics are due to inadequacies in the scientific method, or the basic theories, or both. This writer holds with the latter view.

The science of physics has had ample time to prove its case for the purely quantitative approach to real truth. Unless it can soon begin to answer such questions as: why does the photon always travel at the speed of light in a vacuum? why are mass and energy interchangeable? why does the geometry of space appear to be warped? why does unit charge exist? and why do the attractive forces exist? the credibility of its basic tenets will suffer. Further, it must do this within the context of a unified theory, lest it yield ground to qualitative speculation.

All of this brings to question the validity of the scientific methods now practiced by physicists. The suggestion is not that quantitative investigation is no longer appropriate, but that by itself it is incapable of dealing with matters of fundamental cause. In other words, the complete understanding of the laws and structures of physics might prove elusive until placed within the framework of a qualitative conceptual outline.

Qualitative investigations about the nature of reality lie within the domain of the philosopher. However, philosophy, like physics, has also failed to provide a unified view of reality. Philosophers can't explain why a photon always travels at the speed of light any more than physicists can. Philosophers seem to be as guilty of ignoring quantitative findings as physicists are of ignoring qualitative speculation. One cannot help but wonder how a Ph.D. in philosophy, not knowing the difference between a boson and a lepton, can be expected to meaningfully deal with the question of ultimate reality.

If the qualitative reality of the philosopher and the quantitative reality of the physicist are two different realities, then the two groups may remain aloof from each other. The philosopher can avoid burdensome investigations into the nature of fundamental particles and the physicist can continue to scoff at qualitative cosmologies. However, if both are dealing with the same reality, and both seek to know the entire nature of that reality, then at some point philosophers must put greater emphasis on quantitative investigation and physicists must begin speculating about qualitative cosmologies on which to premise the causes of their observations.

This suggestion is, of course, not new. Indeed, it has only been in the last few centuries that philosophy and physics have been separately regarded. In recent decades there has been a modest, but significant, effort to combine the perspective of both studies in the search for a better understanding of our universe.

Albert Einstein, formulator of special relativity and general relativity theory, is regarded to have been both a philosopher and a physicist. More precisely, he was a theoretical physicist. Theoretical physics is the branch of physics assigned the task of speculating about the nature of the fundamental aspects of physics. Theoretical physicists are, however, denied the philosopher's tool of qualitative speculation; their work is expected to be purely quantitative. Even here the premise that quantitative investigation is sufficient holds firmly and qualitative work cannot get serious attention.

Philosophy, as a study, has been somewhat more successful in approaching the matter of renewing the association between the two investigative fields. With its branch, the philosophy of science, it qualitatively evaluates the methodologies used by scientists. This school of thought studies scientific language, the logic of scientific theories, and other things which concern the way in which scientists conduct their investigations. As a rule, it does not attempt to make comprehensive statements of its own about the nature of reality. This school, for example, would not attempt by itself to offer a unification theory for the four attractive forces in nature.

It is not, even now, unusual to find theoretical physicists in the halls of philosophy of science departments or giving speeches at their conferences. While theoretical physics is not quite as receptive to philosophy yet, the ground of common interest between theoretical physics and the philosophy of science is not unattended. From this ground is arising a new school of thought which attempts to place the observations of physics within some ordered philosophical perspective. The efforts in this area tend to be very individualistic and not easily categorized. Because they arise from the no-man's-land between philosophy and physics, they have difficulty getting serious consideration in either camp. Inasmuch as I cannot find a previously assigned collective name for these efforts, I will call this school: "The Study of Physical Philosophy."

All works in physical philosophy seem to have the three following premises in common:

1. There is but one physical reality which has laws and structures existing independently of human opinion.

2. The capabilities of human observation and human logic are sufficient tools for the adequate understanding of that reality.

3. Direct and indirect observations and experiments are, alone, incapable of exploring the entire spectrum of reality in that reality is both qualitative and quantitative.

This last premise rests on two arguments. The first of the arguments turns on certain implications of the uncertainty principle. Here it is suggested that the observations of the fundamental laws and structures of reality are limited because their investigation can only be conducted by apparatus governed by these same laws and structures. For instance: suppose, in reality, "energy" is but positive and negative units of existence always traveling at the speed of light and that a photon is comprised of millions of these units. Such a fact could never be determined by direct observation since a photon, the most fundamental unit of observation, can never be used to measure anything smaller than itself. The second argument turns on the observed shortcomings of physics already alluded to.

From the three premises of Physical Philosophy arises its basic theorem:

> Reality must be both qualitatively and quantitatively defined in order to be understood.

This means that the study of Physical Philosophy must have its own methods of investigation, ones that are both philosophical and scientific, both qualitative and quantitative. It is the purpose of this book to offer such a method of investigation, a method by which the nature of physical reality can come to be better understood. I call this method of investigation "Consequentialism."

## CHAPTER TWO

# Consequentialism as a Philosophy

Consequentialism, as its name implies, is a deterministic methodology. Yet, its place in philosophy is not under the banner of determinism. Determinism has, historically, been confined to the study of event causation and the nature of personal will. Its perspective is usually from the continuum of time. It does not tend to concern itself with the simple to complex continuum of physical structures. For example, a determinist would not be inclined to explore the relationship between particles and atoms.

Consequentialists deal with both the array of physical structures and the continuum of events. They are as much concerned with the cause of gravitation as they are with the cause of selection determination. It is the belief of consequentialists that all cause and effect relationships, whether physical (particles cause atoms) or temporal (event A causes event B), arise from one common reality continuum. Thus, consequentialism is not traditional determinism because of its scope and also, because of its unique methodology. Indeed, the methodology is so unique that the taxonomy of philosophy requires the study be given its own name; hence: consequentialism.

Any attempt here to trace the history of the philosophical theories and scientific viewpoints which serve as the ancestors of consequentialism would, at best, be very incomplete. There is not room in this book to deal with the extensive array of opinion that forms the backdrop for this philosophy. Like other modern methodologies, consequentialism draws its ideas from more than one school of thought. In addition to determinism one can see in its viewpoints aspects of scholasticism, materialism, absolutism, fatalism, dialecticism, empiricism, and rationalism. To define the views of this philosophy as any one of these things would be very misleading, and yet, consequentialism is not a mere electicism of these philosophies either. Consequentialism is best understood when independently viewed.

As an offspring of physical philosophy, consequentialism shares its three basic premises and one theorem. In addition, it offers a fourth premise which may be stated as follows:

> The laws and structures of the one physical reality are predicated upon one basic, fixed, and comparatively simple set of irreducible laws and structures.

Its individuality as a study stems not so much from its basic tenets, these are not new, but from its methodology. Yet, its methodology follows from the premises.

If there is but one physical reality; and if that reality is founded on a fundamental set of laws and structures; and if those laws and structures are fixed, then it follows that physical reality exists as a definable and predictable system of relationships. It further follows that the complexities of the physical universe must exist as the natural and fixed ramifications of the basic set of laws and structures.

If this is the case, then there can exist only one possible set of fundamental laws and structures which give us our observed universe. Thus, the consequentialist sees reality as a defined system whose unobservable parts can be deduced from those things which can be observed. The methodology of the consequentialist, then, turns on the building of reality models with the tool of logic.

To understand this concept better, consider an imaginary, purely material universe whose only structure is an irreducible homogeneous essence, property, presence, or substance which we will call x. Let us further suggest that the intrinsic nature of this x requires that it always exist as a mirror of itself; much like one could say the letter x is the mirror image of a v. This aspect of the x would function as a law. We will call this aspect the "law of mirrored structure" and state it as follows: "For every structure there is an identical structure."

This universe would express its structure and law in this fashion:

1. x
2. xx

Now suppose that the double x by being distinctive from x alone becomes, itself, a new structure (xx). The law of mirrored structure would require another (xx) and so on as below.

3. (xx) (xx)
4. [(xx) (xx)] [(xx) (xx)]
5. {[(xx) (xx)] [(xx) (xx)]} {[(xx) (xx)] [(xx) (xx)]}
6. etc.

Of course, we are not suggesting the real universe is this simplistic but it may have the kind of innate tautology which allows a basic set of laws and structures to provide us with the diversified universe we know. A computer, for instance, can create a distinct symbol for every aspect of reality we can think of with the possible patterns of only two basic bits.

In this work we will take the position that human logic is nothing more than an ability to recognize patterns.

Pattern recognition allows us to make predictions. Therefore, we will suggest that something may be regarded as logical if it lends itself to predictability. This view of logic is sufficient for its use in this book and forces anyone wishing to deny human logic to also deny the human ability to make predictions. Column A, below, may be regarded as more logical than Column B simply because its next term has predictability while that of B has little.

| Column A | Column B |
|---|---|
| 1. 0 | 1. 00 |
| 2. 00 | 2. 00000 |
| 3. 000 | 3. 00 |
| 4. 0000 | 4. 0 |
| 5. ? | 5. ? |

Of course, line 5 of Column A can only be truly predicted if that column is sequential and deterministic. Yet, how can I know that the number of circles that will first occur to most readers as belonging there is five? If reality is consequential and deterministic, then it has pattern and its various aspects have sequential linkage or pattern. This means that the physical impulses or energy waves which impinge on our five senses owe what pattern they have to the reality pattern.

If logic is pattern recognition, and if the test of logic is predictability, then it follows that the view of reality which is best at predicting future events or yet unseen structures, is the view most in tune with reality. A reality model is essentially an outline of one's view of the spectrum of reality.

If reality is one linked spectrum from the fundamental to the complex, then it follows that a proper reality model must have a similar kind of internal deterministic linkage. The belief that such linkage exists in reality is behind the belief that the laws of physics should be unified.

Philosophers are divided in their viewpoints on physics. Some treat physics as a separate but equal study, quantitatively investigating the universe, while they themselves qualitatively investigate it. Others take the position that physics is only an investigative study and need not concern itself with questions about the nature of reality as a whole. To them, physics, chemistry, astronomy, geology, biology, and the behavioral sciences are merely the eyes and ears of philosophy. The philosopher, then, is to gather all of the data and make some sense of it.

Consequentialism is forced by its tenet that reality is a single system, to hold the latter perspective. Yet, it acknowledges the quantitative work of physics to be especially crucial in the construction of a viable reality model. Further, consequentialism takes the scientific viewpoint, often ignored by philosophers, that no tenet is meaningful unless its logical implications can be empirically verified.

This, then, raises the question of the requirements of consequentialism. What must its methodology deal with? Traditionally, reality has been divided into three areas of study. The first concerns the question of primary cause and the nature of all that is truly fundamental. The term for this area of study is "cosmogony." The second is the area which studies those things which can be observed by the five senses. This area we shall call "matterology," since the term "The Natural Sciences" lacks a philosophical ring. Finally, there is the study of being and feeling. This area is called "ontology." Consequentialism recognizes and utilizes these three divisions of reality, with the exception that cosmogonic speculations are treated under the general heading of cosmology.

The consequentialist may develop models which treat one, two, or all three of these areas. The mystical overtones acquired by the metaphysical studies of cosmology and ontology in recent centuries might cause some to ignore them. However, modern efforts in astrophysics and the neurological sciences are restoring an air of scientific respectability to the questions of origin and personal awareness.

The philosophical studies of aesthetics, ethics, and politics may also be included as a fourth area for a consequentialistic model. However, this area, like the ontology, will lack the test of empirical verification possessed by the cosmology and matterology. Therefore, these two areas might be developed by the consequentialistic model builder, more to define for himself his own views of the nature of these aspects of reality, than for public presentation. For the sake of taxonomy I will refer to aesthetics, ethics, and politics collectively as conductology.

There is nothing in consequentialism that mandates atheism or agnosticism. The consequentialist may be either a materialist or a dualist, one who believes in both a universe of matter and spirit. It is difficult to imagine how a pure spiritualist, one who rejects the existence of matter, could deal with the basic premises of physical philosophy. Most dualists may choose to omit the spiritual aspects of their perceived universe from their presented reality model, judging such things to be beyond model encapsulization. However, if spiritual considerations are incorporated into the model, then the question of linkage between material and spirit should be dealt with.

For example, it is observed that some chemicals affect human behavior and actions. If there exists in the human body a soul also capable of influencing the actions of that body, then what are the mechanics of that influence? How can the atoms of the nervous system, which are known to be subject to the laws of physics, also be subject to another set of laws arising from a non-physical source? Any consequentialist, incorporating a soul in his reality model, must deal with such questions of influence and linkage.

The ancient view that interpretations of reality must always exist as approximations does not compromise man's need for understanding. Entities such as electrons and photons, or their sub-structures, if they have any, may forever elude "perfect" definition. We tend to define physical structures in terms of their constituent parts and certainly the irreducible properties of reality cannot be defined in terms of anything simpler. For example, try to physically describe energy.

Yet, we can show relationships. Even if such things as energy and empty space can never be substantively defined, we can establish how they fit into the general "pattern" of things and make predictions accordingly.

Consider the color red. It can be argued that red as a color is a purely ontological thing. That is, it could be said that red is an awareness of an electromagnetic wave striking the eye with a frequency of around $4 \times 10^{14}$ Hz. Redness, as a color, has no wave length, only frequency. If the light source and the eye have little motion, and if the light is traveling through an approximate vacuum, then an electromagnetic wave length from 650 to 700 nanometers will strike the eye with about the above-mentioned frequency.

If all life ceased in the universe then the color red would cease, though of course (I'm no solipsist), light waves would continue to exist. There would still be electromagnetic wave patterns with wave length, and even frequency, but this is all. For red, wave length is the basic pattern, frequency is that pattern plus the factor of a receiver whose situation is altered because the wave exists, and color is the neurological reaction to this secondary pattern. Can we be sure that an insect sees the color red just as we do? How do we know that its simpler neurology does not cause it to see red as we see yellow, for instance?

Any matterology seeking to describe electromagnetic waves can only speak of pattern. If the photon is indivisible, we cannot speak substantively about light waves. That would be like asking: What is energy made of? The origin of the photon, or energy waves, would have to be dealt with in some sort of cosmogony. The cosmogony, then, of a reality model must provide us with a set of fundamental and irreducible "set of parts" for our universe; the matterology would describe the pattern or potential relationships such parts have; and the ontology would link our perceptions to those patterns.

Since perception is linked via pattern to the fundamental structures from which existence is formed, correct pattern recognition could provide us with a total view of reality if that reality is consequentialistic. Figure 1 portrays this notion diagrammatically.

The methodology of consequentialism assumes that the ramifications of our universe are based upon one fundamental set of truths. This perspective finds it very difficult, therefore, to separate a given cosmogonic origin theory from whatever cosmology it implies. Thus, from this point onward the terms cosmology and genesis model will be used interchangeably to imply both an origin theory and its attendant structuring of the universe.

## The Epistemology of Consequentialism

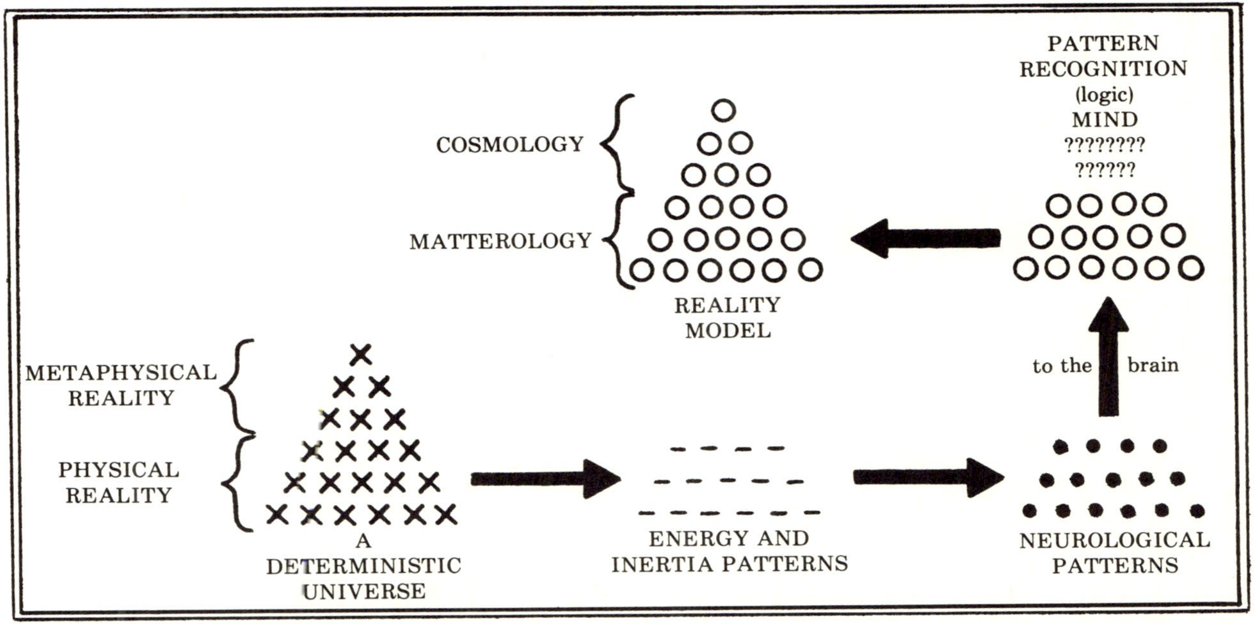


Figure 1. Consequentialism takes the view that our sensory perceptions are patterns which owe their arrangements and character to a tautological reality. If reality is sequential and deterministic, and if our neurological patterns are functional to that reality, it follows that through the brain's (mind's, if you prefer) ability to recognize pattern the entire scheme of the universe's tautology can be deduced. However, the precise essence or substance of reality may forever escape our understanding since that cannot be known with pattern recognition.

Consequentialism, in its entire form, is more than just a methodology; it is a complete philosophical system, possibly capable of dealing with all aspects of reality. If the philosopher can deal with the question of primary cause, present a physics which conforms with the observed universe, develop an explanation, even in general terms, for personal awareness, and, all within a unified set of laws, conditions, and structures, then he can enjoy the sense of having a complete understanding of his universe. Further, from such an understanding he can develop a system of aesthetics, ethics, and politics which conform to his view of the universe.

## THE SUMMARY OF CONSEQUENTIALISM AS A PHILOSOPHY

Consequentialism begins with four assumptions about reality:

### The Premises of Consequentialism

1. There is but one physical reality which has laws and structures existing independently of human opinion.
2. The capabilities of human observation and human logic are sufficient tools for the understanding of that reality.
3. Reality is both qualitative and quantitative.
4. The observed laws and structures of the one physical reality are predicated upon one basic, fixed, and comparatively simple set of irreducible laws and structures.

From the assumptions arise the following theorems:

**The Theorems of Consequentialism**

1. Reality must be both qualitatively and quantitatively defined in order to be understood.
2. Reality is a definable continuum of cause and effect relationships.
3. By properly defining the continuum, the unobservable parts of reality can be deduced from those parts which can be observed.

From the theorems arises a three-part method of inductive investigation which qualitatively and quantitatively analyzes the logical ramifications of various sets of irreducible laws and structures, seeking that set whose implications best fit our observed universe. The first part of this method is purely qualitative, the second part is both qualitative and quantitative, and the third part is purely quantitative. (The methodology will be explained in the next chapter.)

From the methodology arises a two-part reality model which attempts to define the continuum of cause and effect relationships which is physical reality. The first part of the model is a cosmology which explains how the physical reality came to be and the second part is a matterology which explains the observations of the universe in terms of the cosmology.

The theorist, if so choosing, may develop an ontology from his cosmology and matterology with which he will attempt to place being and personal awareness in his reality model.

As a final option, the theorist might attempt an evaluation of what should constitute proper conduct and standards of values for himself and others in light of his model of reality and implied ontology. This evaluation, which we call a conductology, would place aesthetics, ethics, and even politics within the context of the reality model.

The ultimate goal of a consequentialistic model builder is a four-part reality model having no internal contradictions, complete fit with the observations of the real world, and internal linkage which establishes a cause and effect relationship from implication to implication. Are such comprehensive models possible? It will be a function of this book to show that they are.

## CHAPTER THREE

# Consequentialism as a Scientific Method

When the premise is held that all of the complexities of the universe are consequential to a comparatively simple set of laws and structures, it is a fair question to ask: Why were not the mysteries of physics solved long ago? In light of the tremendous investment of time and talent into research, it seems logical to assert that if such things as gravitation, particle decay, and the cause of permanent magnetism were indeed simple, they would be fully understood by now.

However, this argument can only be held on the grounds that all of the basic assumptions of modern physics are valid. It is not outside the bounds of reason to suggest that the large number of unsolved mysteries present in modern physics are there because certain fundamental tenets of the science are either inadequate or inaccurate. In fact, anyone seeking to simplify the structures and laws of physics must hold that this is indeed the case, lest he insult the intelligence of the community of physicists. Since, historically, many of the most brilliant minds this world has ever produced have been members of this community, it must be held that if the basics of physics are simple, then certain fundamental assumptions in physics are inaccurate. The question, then, that must be considered is this: What evidence is there that a simple set of laws and structures do indeed underlie the complexities of modern science? By simple, I mean simple in the popular sense.

Firstly, there is evidence that the most basic structures of the universe may be simple in number. When molecules, atoms, particles, and quarks are considered as categories, the known classes in each category and the complexity of the structure are respective. That is: the simpler the structure, the fewer kinds there are of it. While there are more atoms than molecules in the universe, there are fewer types of atoms than types of molecules.

Perhaps a clarification is needed at this point. While the number of kinds of particles exceeds the number of kinds of naturally occurring elements, the number of stable particles does not. If atoms and anti-atoms were constructed from the list of known particles, the memberships of possible elements, possible molecules, and possible compounds would grow accordingly. Thus, it is valid to say that the simpler the structure, the smaller the number of classes in its category.

When the number of kinds of quarks, the number of kinds of leptons, the number of kinds of bosons, and the number of fundamental properties are considered, it is evident that the universe, even in modern theory, rests on a comparatively limited set of laws and structures. Further, it is possible that these now fundamental entities are not fundamental at all but exist as the natural implication of an even simpler set of laws and structures. Allowing for the need for symmetry suggested by charge, spin, and the other mirrored properties of particles and antiparticles, it is likely that the number of fundamental divisions of reality cannot be less than two.

If we, perhaps arbitrarily, set the number of quarks and antiquarks at twelve; the number of leptons and antileptons at six; the number of neutrinos and antineutrinos at six; and the number of bosons at six;

then it can be seen that any cosmology or model of the basic structures of the universe would not have more than thirty units or less than two.

Secondly, the descriptive properties of what are now regarded as fundamental structures are few in number. An individual electron, for example, has only velocity, charge, mass, magnetic moment, and spin. Outside of the additional characteristic of lifetime, not much else can be said about its physical description.

Thirdly, there is simplicity in the known laws of physics. Newtonian mechanics, the law of gravitation, basic atomic structure, and the basics of electromagnetic field theory are easily understood with only a modest background in physics. For all of its seeming ramifications, the mass-energy relationship of special relativity is ultimately reducible to the simple formula of $E=mc^2$. Even quantum mechanics needs only integers or simple fractions to express such concepts as particle spin and charge.

That physics has a history of expressing some of its most profound truths in simple terms may be one of the least appreciated facts in all of modern-day science. For example, most of the work now being done in gravitational theory is a graduate student's nightmare of tensors, Hamiltonians, orthonormal frames, and other esoteric mathematics. Yet, when the understanding of the actual cause of gravitation is finally reached, it will probably be readily understood, in principle at least, by the average ninth grader. There have been thousands of authoritative papers produced to date on the subject of the cause of gravitation, yet it still remains a mystery. It is not impossible that this arcanum will finally be solved by the humble reasoning of a single individual.

From these three arguments it is established that the human search for a simple set of fundamental truths to explain the basics of the universe may not be a fruitless one.

In discussing the evidence supporting the notion that the fundamentals of physics are simple, it was pointed out that the membership numbers of each category of matter lessen with respect to decreasing complexity. This fact can be conceptualized as a cone of categories.

**The Cone of Categories**

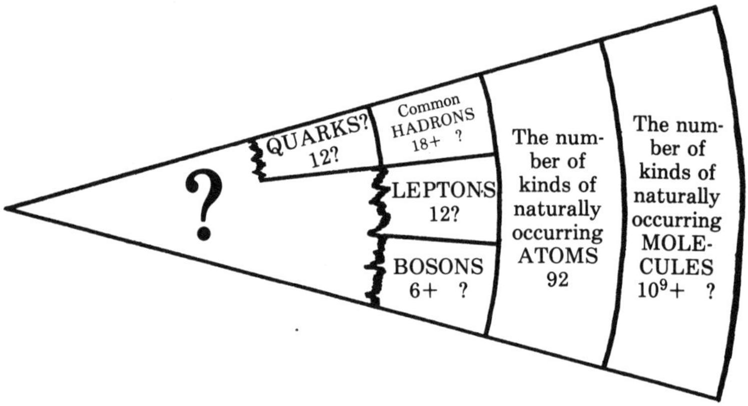


Figure 2. The basic structures of our physical world tend to decrease in family members as they decrease in complexity.

In general terms, it can be said that the actual discovery of the existence of each category of matter came from the study of more complex structures. Man learned of compounds and molecules by studying everyday objects. In spite of the antiquity of atomistic theory, the initial confirmation that atoms exist came from studying molecules. Likewise, particles have been discovered from studying atoms, and quark theory has been derived from particle studies. Thus, the generalization can be made that the thrust of scientific research, within the context of the cone of categories, has been from right to left, from the more complex to the less complex.

Scientists, using scores of laboratories and spending millions of dollars, are working around the world to answer the questions that the builder of any consequentialistic model must deal with. It would be an overwhelming task for any single individual to try to approach the question of fundamental structures along the heavily-trod path just described. What is needed is an original approach to the mysteries that lie at the tip of the cone of categories. Therefore, consequentialistic methodology approaches the question of origin by examining the cone of categories from its tip end.

**Two Investigative Approaches**

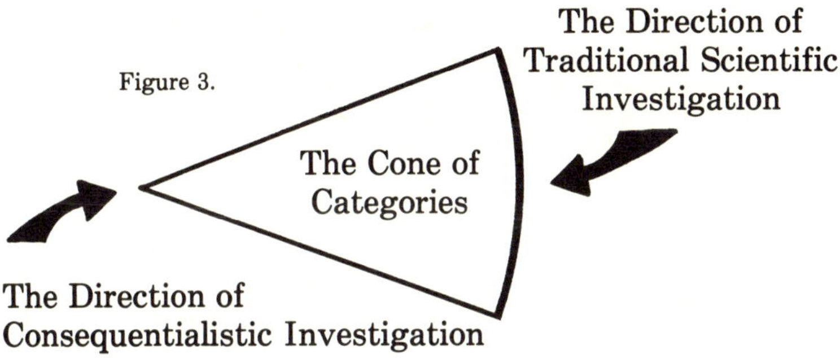


Figure 3.

It is this change in investigative direction that distinguishes the methodology of consequentialism. By suggesting that the basic ingredients of physical reality are few in number, the implication is made that the number of possible sets of fundamental laws and structures which can be theorized are not unmanageable.

This opens the possibility that through enough logical inductive elaborations on various hypothetical structures and laws, a set might be found whose logical implications fit our observed universe. If the fit is good enough, the metaphysical origin theory mandated by the hypothetical set could be wed with the physics of observed reality, allowing new advances toward the goal of a total understanding of our universe.

The first step of this scientific method then must have two phases: a first phase which postulates an irreducible set of laws and structures and a second phase which analyzes their intra-relationship in purely qualitative terms, while outlining the implications of those relationships. If the qualitative implications, based on arm-chair logic, do not yield a physics which provides a prima facie fit with the observed universe, then certainly any quantitative investigation would be fruitless. Hence, the qualitative review of the postulates acts as an efficient screening device, freeing the theorist to review a wide selection of postulated cosmology models, unburdened by the time-consuming task of quantitative study. Here, then, are the investigative steps of consequentialism.

## STEP ONE IN CONSEQUENTIALISTIC METHODOLOGY

As already noted, the first thing a consequentialist must deal with is the question of first cause. He must postulate a prime mover or (borrowing from determinism) a primary sufficient condition. His first cause might be a defined void, energy, time, empty space, a Creator, or even the structures and laws now thought to be basic in physics. Whatever his postulated starting point, it must be credibly argued to be irreducible.

His next job is to examine his postulated primary sufficient condition for its consequences. For example, suppose the theorist postulates his prime cause to be a defined void comprised only of energy, anti-energy, empty space, and time. He must then postulate a set of relationships between them which in turn give rise to even more complex relationships. The final outcome of his postulated relationships and ensuing structures will be a set of observable laws and properties. This completes his cosmology.

To this point the theorist is completely free in what he says. His universe can be anything he postulates it to be. His only requirement is that the postulates must be logically related. He can define such things as empty space, void, volume, distance, time, and energy in any way he wants. If he postulates that empty space, void, and unconfined volume are all the same thing, that is fine. If he says energy and time are the same thing . . . so what. Postulates cannot be attacked for what they say. However, their logical implications must yield verifiable results. The cosmology can only be challenged on the basis of what it implies.

Because this part of the methodology is inductive, the theorist needs some way for screening his cosmology. The likelihood of his picking just the right combination of fundamental laws and structures and the right sequence of interactions on the first, or even fiftieth, attempt is quite low. As will be seen, steps two and three will be very quantitative and in line with traditional scientific method. However, quantitative verification is a very time-consuming task. The theorist wants to be able to judge his cosmology model quickly for its fit with the observed universe. In short, what he needs in this first step is a qualitative method of review which will check for potential quantitative fit.

Phase two of step one thus requires the theorist to place the logical ramification of his cosmology within the context of the direct observations of physics. Here are some typical questions the theorist should ask himself.

Do the logical implications of the model account for:

The mass energy-relationship?

The existence of protons and electrons?

The typical modes of particle decay?

Beta decay and neutrino phenomena?

The four attractive forces in nature?

Electromagnetic wave propagation?

The Young interference experiment?

The refraction and diffraction of light?

The bending of light by the sun?

The slowing of Mercury?

If the logical extension of the cosmology cannot offer the theorist common sense explanations for these basic phenomena of physics, then it is probably incapable of serving as a reality model. Of course, the argument can be made that a model might not be able to approach the fundamentals of physics from a common sense qualitative perspective. Indeed, much of modern theory rests on the rejection of the common sense view of our universe. This question comes to the very heart of consequentialism.

First of all, to the consequentialist, common sense and logical order are the same thing. To him, the common sense view is not the traditional ideas of the masses, but ordered explanations which are in line with direct observation. Therefore, he might view special relativity as a common sense theory while rejecting the uncertainty principle as an illogical explanation for the closed-window Young interference experiment.

Two of the important foundations of consequentialism are that the universe is simple in terms of primary structure, and that the human mind is capable of understanding that structure and its ramifications. The reasoning for this latter tenet goes this way: if the biological mind is the product of an ordered universe, and if all sensory perceptions are the perceived patterns of that same ordered reality, then to the consequentialists, the human mind rests within the confines of that order and is incapable of escaping pattern recognition.

We are all as blind men with an elephant of reality. The fact that we can recognize and compare patterns does not alone guarantee that we will come to know all of reality. However, it does assure us that, if given enough parts of the puzzle, we could put it all together with the tool of pattern comparison. Logic is nothing more than comparing one pattern with another. Had one of the blind men been a consequentialist he, upon hearing the comments of the others, could have properly described the elephant.

Any argument that the questions of physics cannot be approached with a common sense ordering of things rings empty in the ears of the consequentialist. Indeed, he interprets the paucity of common sense explanations for the observations of fundamental physics as evidence that many popular theories are not adequately interpreting reality.

To prove that such common sense qualitative ordering is possible for a good cosmology, this book offers a sample of this first step in the second section of this book. The sample reality model, offered here, was developed by this theorist as a way of proving this methodology to be viable. It first outlines an origin theory for the universe which results in three basic structures: they are empty space, a positive unit of influence, and a negative unit of influence. The two units of influence, called vectons, have a set of personal properties and a set of rules governing their interactions with each other. This constitutes the cosmology of the vecton model.

The potential fit of the sample model with observed reality is then approached with each of the questions already mentioned being dealt with, as well as many others. What emerges is a prima facie fit with the key observations of physics. In other words, there has been nothing actually observed which says the universe might not exist exactly as the sample model says it does.

The qualitative exploration for fit with the experiments and observations of physics can form the basis for a model's matterology. It is not essential that any new matterology agree with the popular interpretations of these experiments and observations. However, it is important that when a contradiction between popular interpretation and the implications of the model occurs, that the model offer a logical explanation for the experimental results. For example, if a model suggests that the missing energy of a beta decay does not manifest itself as a neutrino, then some explanation should be given for those experiments in which neutrinos were supposedly observed.

In the vecton model offered in this book, several such discrepancies occur. The model, for instance, finds that all photons are slowed, whether passing near a gravitating body or not, but that the degree of slowing is proportional to the photon's distance from mass. Hence, all photons are probabilistically slowed as they travel through empty space in direct proportion to the distance they cover. Thus, the Hubble red shift and the red shift of light passing near the sun arise from the same basic cause according to the model.

The theorist must be able to distinguish between those things which have been actually observed and those things which are popularly believed. For instance, the universe has never been observed to be continuously expanding; this is only a theory based on the assumption that the Hubble red shift is a Doppler shift. The only true observation is the red shift itself. The theorist must be careful to deal only with those things which have been actually observed in establishing the qualitative matterology implied by his cosmology. Popularity is no foundation for truth.

## STEP TWO IN CONSEQUENTIALISTIC METHODOLOGY

It is not likely that the community of physicists will give much serious consideration to qualitatively presented ideas. Therefore, the theorist, no matter how exciting and seemingly successful he finds his reality model to be, should recognize that he, with step one of this method, has only a cosmology and not a theory in physics. However, every theory needs a qualitative conceptual outline or set of premises as its causative foundation. It is the job of step two to begin to place the qualitative ideas of the cosmology on some kind of quantitative footing so that the ideas of the theorist can at least receive academic review.

As already established with the screening technique of step one, every cosmology, upon logical extension, will make certain inescapable predictions about reality. For a good model most of these predictions will be found in other theories as well, but some will be unique. If one or more of the unique ramifications of the cosmology are placed within a quantitative framework, then an experiment can be set up under the rules of conventional scientific method to test the quantitative predictions of the qualitatively arrived at concept.

If there is agreement between the cosmology prediction and the experimental results, then of course, the cosmology and its attendant quantitative research can be set before the scientific community as theory. Step two differs little from conventional scientific method other than in the source of the premise. Most theories studied now are the mathematically arrived at implications of other theories, usually the major three previously referred to. In the model method conformity with popular theory is not sought. The studied prediction, while quantitatively formulated, has its roots in an independent qualitative cosmology and not in other theories.

Many college textbooks in physics present the ideas of popular theories as fact. Since today's physicist was yesterday's student, people in the science often confuse that which is popularly believed with that which has been actually observed. For example: Everyone knows that mass cannot travel at the speed of light. Therefore, any theory offering a unit of mass traveling at that speed must be in error. However, observation has only validated this statement for particles and structures made up of particles. It is not impossible that a subparticle unit of mass might exist which travels at the speed of light. Hence, the indisputable truth that everybody knows may not be a universal one.

It must be an intrinsic part of the model method of investigation that its tenets not be put in the light of popular theories. Its conformities need only be with that which has been actually observed. Indeed, the lack of progress in the unification of the laws of physics may stem from the insistence by established scientists that new theories be quantitatively compared with the popularly held old ones.

A second unique feature of the cosmology — implication — experiment sequence is that the experiment always follows the theory. This modest distinction could dramatically alter the investigator's perspective. Both the theory of special relativity and the uncertainty principle, for example, are based on experiments conducted before the concepts were promulgated. The experiment — theory sequence tends to cause the theory to be quantitatively expressed from the viewpoint of the observer. This observer perspective is so ingrained in physics that few ever stop to consider its validity in defining reality.

Consider two photons traveling exactly head-on in a vacuum. The observer perspective denies them a relative speed of twice the speed of light despite the common-sense view that is indeed their relationship. Since the mathematical constructs of special relativity are predicated upon the observer perspective, the common-sense perspective is supplanted by the mathematical "proof." However, the degree to which this so-called proof reflects the real relationship between the two photons can only be determined by application of the theory to other experiments. So far all experiments hold that the speed of light is the most upper bound for absolute speed, but no physicist can point to any experiment which proves that it *is* the most upper bound for relative speed. That idea is only held as a theory and may be completely out of touch with reality even though mathematically precise.

Many physicists reject logic and qualitative reasoning in favor of mathematical proof without realizing that these two mental processes are the parents of mathematics. The contemporary notion of infinity, for example, is not only qualitatively defined but also mathematically used solely from the experimenter perspective.

To understand this argument better, consider a point A radiating an infinite number of straight lines. Now consider a second point B, not A, also radiating an infinite number of straight lines. We will suggest that the cardinality of straight lines radiating from each point corresponds with the number of directions each point may express so that both straight line sets may be regarded as equally defined.

From the experimenter point of view the number of straight lines involved always has the same value whether the points are considered in summation or individually. Infinity once is equal to infinity twice because the experimenter can't count one infinity any more than he can count two. Hence, the experimenter perspective brings to mathematics the idea that infinities cannot be relative. However, the operative laws of reality may indeed be predicated upon relationships between infinite quantities so that, in the above example, the statement that point A radiates half as many straight lines as points A plus B combined is a valid one.

The point is this: both the classically defined infinity and the relative infinity ultimately rest upon qualitatively arrived at arguments. Similar qualitative foundations exist for the definition of zero, irrational numbers, and other aspects of mathematics. The model method of scientific investigation, by relying only on direct observation and not on popular theory, forces a constant reevaluation of the qualitative tenets of mathematics and physics. Such reevaluation is essential if the science is ever to distinguish between those things it believes because they have been observed and those things it believes simply because they have always been believed.

One question that arises with step two is this: How can a qualitative cosmology yield quantitative predictions on which to premise quantitative experiments? The answer to this lies in the logical implications of the cosmology which serve both as its screening device and matterology.

A good qualitative cosmology is not some mystical poetic thing. Rather it sets precise rules regarding quantity and interaction. Ideally it should even create its own mathematics. It should address such questions as whether zero is the sum of two equal opposites or simply nothingness, and whether or not some things can be more infinite than others. Logical order and consistency are its only real criteria in its initial draft.

Here again the sample model in Section II will be instructive. Its vectons form particles by forming circular force lines. The force line sets are called vecton clouds. However, since only one kind of cloud is formed, the model must argue that quarks are lepton-like structures. It then follows that leptons and the charged pion have one cloud each, the neutral pion and mesons have two clouds each, and baryons have three or more clouds. It is another prediction of the model that all neutral particles must have an even number of clouds. The implication is provable because all observed typical modes of decay preserve cloud number just as the model implies. The defined quantities and defined interactions of the model make defined quantitative experiments possible.

A good reality model should make one entirely unexpected prediction which is verifiable. The vecton model's explanation for gravitation, for example, predicts that all weak force decays randomly scatter free vectons which have gravitational influence. Since random scattering always yields an influence inversely proportional to the square of the distance from the source, the vecton model can predict that all weak force decays yield a gravitational influence which is directly proportional to the energy released and inversely proportional to the square of the distance from the point of release. Qualitative cosmologies can indeed yield verifiable statements.

Most theorists will be unable to conduct the test or tests of step two for themselves. Most rigorous experiments require apparatus not readily available. Therefore, if the theorist does have a cosmology whose logical extensions provide a fit with observed reality (step 1), and if those logical extensions do offer a unique and verifiable quantitative prediction (step 2), then the model may assume the title of theory and be presented to the scientific community for verification through traditional testing techniques (step 3).

## STEP THREE IN CONSEQUENTIALISTIC METHODOLOGY

Step three follows the traditional scientific method completely. Here the quantitative implications developed in step two are investigated through actual experiments. Hopefully, step three will be initiated by an experiment which can prove a unique prediction of the reality model. However, certain models may not be able to offer new predictions and must rely on their fit with existing observations. If this is the case, then the argument for the model must be held on the grounds of its incontestability.

Whether the initial fit of the model lies with a successful experimental prediction or a good fit with the classical experiments of physics, such compatibility alone does not prove the theory. It will, however, justify further investigations into other predictions of the model from which may arise refinements in the matterology, or even the cosmology.

## SUMMARY OF CONSEQUENTIALISM AS A SCIENTIFIC METHOD

### Step One — Purely Qualitative

1. Develop a well-defined set of postulated laws and/or structures which might serve as the causative factors for "the" universe.
2. Study the logical implications of the postulates to determine if their extended ramifications yield a prima facie fit with those things in our universe which have been actually observed.
3. If such a fit exists, develop a matterology which qualitatively defines the observed physical universe in terms of the postulated cosmology.

### Step Two — Qualitative/Quantitative

Begin the transition from qualitative cosmology to quantitative theory by selecting one of the implications of the reality model which makes a unique and verifiable prediction. Put this prediction within a quantitative framework and outline experiments which might prove the prediction.

### Step Three — Purely Quantitative

Conduct rigorous experiments designed to prove or disprove the selected prediction of the theorist. If the model's predictions are verified by a unique experiment conducted with traditional scientific methods, or by a fit with several classical experiments, then the model may be regarded as viable theory and tested further for refinements.

## CHAPTER FOUR

# Developing Reality Models

It is one thing to suggest that a variety of reality models can be developed and then inductively examined for their fit with the observed universe, and quite another to actually construct such models. Therefore, without a model to serve as an example of the consequentialistic approach to understanding reality, this book would be somewhat meaningless. Accordingly, Section II of this book will serve in its entirety as a sample reality model.

Steps two and three of the methodology are not so new that they will need examplification. However, step one is totally unique and hence will command the attention of most of Section II. As the author of the sample model as well as this book, I will be able to not only show you my postulates of the sample's step one, but also share with you the thinking on which they are predicated.

The vecton reality model is probably the best of several I developed. If I had put each of my models through elaborate mathematical and laboratory proofs, I would probably still be working on my first outline. Instead, what I developed was a kind of checklist, a set of observations which I tried to explain in terms of my model. The fits were always qualitative and at best superficial. Nevertheless, they did tend to reveal glaring shortcomings in the model.

Humble reasoning is the theme of this work. Every cosmological model (I like the term genesis model) has a logical "implication" with regard to each observation in physics. For example, we might ask: If vecton theory is true, what would be the nature of particle structures?

The question would then be answered with a generalized narrative explaining the nature of quarks, particles, and particle decay within the context of vectons. The statements in the narrative would be primarily qualitative and unverified. If these statements have plausibility, then they may stand for the time being. Next, a second area of physics can be approached, and so on.

Most genesis models will have difficulty yielding plausible explanations for more than two or three observations. Hence, a checklist, based on nothing more than logical implication, is an excellent way of screening the models (step one, phase two).

For the sample model I have chosen three areas of physics as my checklist. They are particle structure, the attractive forces, and the wave nature of light. With each area some basic implication is suggested about their nature if vectons are a reality. For example, the four attractive forces must somehow arise from negative vectoring and the exchange of vectons.

Using little more than armchair reasoning, I next contemplate the possible fits the implications have with certain selected observations made in each area. From this "personal debate" of the model's ability to explain the observations of physics I am able to develop a matterology as well as outline of the quantitative measures I must make to argue the model.

Because this step is inductive, most models will fail this test of general fit. Prima facie fit, though qualitative, is still not easily achieved. The theorist must then either modify his model or begin with an entirely new one.

If a model can provide a series of logical fits by offering plausible generalizations as a way of explaining certain selected observations in physics, it then approaches its first quantitative hurdle. It must offer a logical implication which cannot be explained in terms of any other theory. It must predict something totally unexpected which is to be its proof of viability. This is the nature of step two.

Section II takes vecton theory as far as step one though we will touch on steps two and three. Our explained implications are at best plausible, but unverified, generalizations about the fit of this genesis model with the observed universe. In step two we will make a quantitative prediction which, if found to be true, might serve as strong evidence that vectons or vecton-like entities exist.

If this, or any other genesis model, is able to pass its quantitative hurdle, then it should be set before the academic community as viable theory to be quantitatively studied. The model will then be subject to either proof or disproof on the meaningful field of quantitative evidence (step three).

Genesis models, like the one being offered here, could serve as important tools in the advancement of theoretical physics. Physics has, since the theory of special relativity, entered to some degree the realm of metaphysics. "Whys" are beginning to emerge whose only answers may lie in some metaphysical cosmology. Why must a photon always travel at the speed of light in a vacuum? or, Why does charge not vary with mass? are only two of the questions which may never be answered in traditional terms.

With plausible genesis models one can provide possible "whys" for some of the aspects of theoretical physics. For example, our vecton theory provides a "why" for such things as absolute speed and the relativistic relationship between motion and time.

In addition, genesis models might offer alternative explanations for some of the observations of science. At the time of this writing the Hubble red shift can only be explained as a Doppler shift. Our vecton theory suggests that inter-galactic red shifts need not be Doppler shifts. Such explanative options could serve to iron out wrinkles not solved by current theory.

Consequentialism is as much philosophy as it is physics. It must be kept in mind that a genesis model is only the cosmology of a potentially larger model of the universe which could encompass all aspects of perceived reality including such intangibles as human interaction. By grouping all such interactions under the general title of conductology, a complete reality model can outline everything ever observed under only the following four titles:

Cosmology — The study of primary structure.
Matterology — The study of all natural sciences.
Ontology — The study of awareness and being.
Conductology — The study of relationships and values.

Usually, individual studies are easily classified under one of these four titles. For example: politics, ethics, aesthetics, economics, and law all obviously come under conductology while geology, chemistry, astronomy, and physics obviously fall under matterology. However, some things, such as communication might involve two or more categories. For instance, the physics of a transmitted message would be studied under matterology, the logic of symbolism might be studied under ontology, while the effectiveness of a given communication could be studied under conductology.

Certain philosophies and theologies have presented such broad approaches to the question of existence that they may be regarded as reality models. As already noted, there seems to exist in man the need for such all encompassing outlines of reality. It may very well be that the longevity and popularity of the Pentateuch, the first five books of the Old Testament, might be due to the fact that it serves as a kind of reality model. Though its matterology is modest and its ontology weak, it does offer an origin theory and a rather extensive conductology.

Except for fundamental physical principles, the natural sciences are well understood at the time of this writing. Therefore, any builder of a reality model need not take his matterology beyond the level of atomic structure. If his model can deal with the observations which relate to particle behavior, the attractive forces, and wave mechanics, the rest should automatically fall in line.

Because genesis models and matterologies can serve as inductive tools for defining our physical universe, we are treating them in this book as methodological steps. But philosophically, like the ontologies and conductologies, they are statements about the nature of reality. Therefore, many individuals may wish to build reality models, not to solve the mysteries of theoretical physics, but to define their own personal perspectives of reality. For this reason I have in Section II added discussions on the ontological and conductological possibilities of the sample reality model.

As suggested in the introduction, a theorist in consequentialism needs a background in mathematics and modern physics in order to realistically approach the task of building a good cosmology and matterology. Assuming the reader has such a background, we will move on to the next requirement of the model builder: he needs to be a skeptic in the popular sense of the word! Skepticism is somewhat meaningless without alternatives, and so the third requirement of the theorist is that he or she be an original thinker. The three requirements of knowledge, skepticism, and original thought are necessary from the very beginning in the construction of a reality model.

The first thing the theorist must do is to relax. He should place himself in an environment where he is free to let his mind contemplate the nature of primary cause. At this point all model builders may class themselves as either creationists (dualists) or materialists (monists).

For a creationist, the question most likely dealt with will not be the cause of God or His nature, but how He might have created the universe. That is, what would be His starting point? The creator in most modern theologies is an ordered God, setting forth an ordered universe. Indeed, many theologians argue that the physical order found in the universe is evidence of an intelligent Creator.

For the dualist there is no need to argue primary cause. God always was and that is all there is to it. For him the question of origin centers around which fundamental laws and structures God chose as the foundation for the universe. As a consequentialist, he believes the complexities of observed reality are the intended ramifications of some basic set of laws and structures. If he believes the decisions of men are part of that same intended plan, he is a predestinationist. If he believes the decisions of men can be influenced by factors outside of the physical reality, then he must deal with the questions of linkage already alluded to.

The creationist might first wish to consider the texts and numerologies of his religious beliefs in his initial contemplations. He should ask himself if the present popular theories offer views which are compatible with those of his faith. If his beliefs allow such contemplations, he might even try to put himself in the role of creator asking: "If I were God, how would I go about making a universe?"

If the theorist is a materialist, his speculations are necessarily more complex. He is initially faced with two alternatives: either he may argue the universe is without beginning, or he must argue that the universe came to be from nothingness. If he takes the position that the properties and particles of physics are eternal into the past, then he must overcome the intuitive notion that all physical things must have origin. However, if he argues origin, then he must deal with the question of a prime mover or primary sufficient condition.

As will be seen, our sample model postulates the universe to have both origin and an eternal past. The universe is postulated to exist as a void whose origin lies in a structural sequence rather than time sequence. This allows the model to present a universe with both an eternal past and a primary sufficient condition. The sample model is a materialistic one.

In these early armchair speculations it is important that the theorist bear in mind that many of the basic tenets of physics are held more on tradition and assumption than direct observation. Physics is full of such traditional assumptions. For example, any line or wave front is assumed to be infinitely divisible. Yet, it can be fairly asked: Is this the case in reality? Similarly, it is assumed that infinity, distance, motion, and direction are derived from the existence of matter and energy. Yet, what if matter and energy are predicated on the prior existence of these "concepts"?

We assume that all space is infinitely divisible. Is this valid? Even though we can mathematically depict this, does this mean that this is the case in reality? What if there is a least possible unit of space, greater than zero, which if made any smaller would equal zero? In other words, is it possible that space itself occurs in quantum units? If such a least volume does exist, then it can be fairly argued that every real direction expresses volume.

Is it possible that the volumeless direction of classical mathematics does not adequately represent reality?

What is time? What is it predicated upon? What if there were a sub-particle structure which always traveled at only the speed of light, even when "in" a particle? If all matter and energy were comprised of such units, would they not be the ultimate cause of time? Would not their random loss from all matter influence the space around matter? Might such units be the cause of gravitation?

The two categories of tangible and intangible begin to blur as we approach the ever-changing border between physics and metaphysics. This leads us to ask: Could it be that what we think of as tangible matter is, in reality, no more tangible than volume, direction, distance, and motion?

It is sometimes very difficult to distinguish between those things we know to be true through direct observation and those things we think are true because everyone around us believes them. Essentially, the building of a genesis model is nothing more than playing a game of "what if". Therefore, the author of the model must be free to challenge the existing interpretations of common and uncommon experience. Indeed, he is obligated to do so if he expects to offer anything new.

In the case of the sample model, I began speculating about how a universe of "something" might come from "nothing". There seems to be no reason why anything, even a God, should exist. Why is not all of reality eternal nothingness? Yet, matter and energy do exist and so I am forced to conclude that some primary sufficient condition either does or did exist to cause the universe we know.

My thoughts then turned to the question of whether or not nothingness itself could serve as some form of prime mover. It wouldn't have to create the observed universe all at once, but simply give rise to one other "thing" (time, perhaps?) which could in turn give rise to something else, and so on, until all of the ingredients of the universe we are familiar with existed. Such is the nature of consequentialistic thinking.

This line of thinking led me to contemplate the nature of nothingness. What is it? Is it the same thing as empty space? If I use the words "void" and "nothingness" interchangeably as one concept, can I be assured that concept has a parallel in reality? If the human mind says a primordial void, by definition, could not possibly be defined or even contain anything, does that really mean anything? Is reality what the dictionary says it is?

There are many popular ideas about what a primordial void would be, if indeed one ever existed. Could such a void express volume and direction? Could it, by its own intrinsic nature, have definition? If it has definition, can it truly be regarded as nothingness?

In classical thought, volume arises as the product of three dimensions which can only be expressed with three distances and three directions. This is so ingrained in our thinking that the idea that volume can be expressed without height, depth, and width seems impossible. What if this notion, probably arrived at from our daily association with objects, is not in keeping with reality? What if volume is expressed by empty space and hence a void? In other words, what if our distinctions between volume and space are entirely artificial? Can we say the universe expresses infinite volume?

From these kinds of speculations I came to several conclusions about the nature of a primordial void. First of all, it is impossible that it could have expressed nothingness in the classical, dictionary or popular sense. At least it would have had the property of empty or unconfined volume, in the sense that unconfined empty volume is unconfined empty space. Secondly, this lack of confinement automatically mandates a natural unit of measure which is infinity. The empty space of the void would be without limit.

It seems intuitive that this infinite unordered empty space would naturally always be empty. One would not expect cars, or rocks, or even protons to suddenly appear in such a void. Therefore, the emptiness of the void must always be "conserved". Hence, if ever there existed a primordial void, then it would have, by its very intrinsic nature a fundamental law of conservation, which would require this infinite unbound volume to always be empty (emptiness conserved).

From this reasoning I concluded that if the universe ever existed as a void, then it would have been required to have had the three properties of volume (space), measure (infinity), and conservation law guaranteed (net value). Thus, the universe could never have existed as nothingness.

The next thing I turned my attention to was this intrinsic law guaranteeing that the void always contain nothing. However, my own reasoning showed me that nothingness is an impossibility. This led me to conclude that a void must always have a guaranteed net value of zero and that zero is not nothing. But if zero is not nothing, then what is it? If zero is not nothing, then it must be the sum of two equal opposites. Hence, empty volume is space with a value of zero!

I say these things, not to argue my model's logic, but to indicate the kind of open thinking one needs to initiate his own cosmology. Let us now turn to my sample model, the Vecton Model of Reality. Starting from the conclusion that any primordial void would be an infinite undefined volume of space having conserved a net value of zero, I will try to explain the logic that led me to my postulates and my matterology. It might be helpful if you treat the sample genesis model, not as a cosmology of our own universe, but as that of a hypothetical universe. In this way you might be able to deal with its original ideas more objectively. This model will show by example, the methodology of consequentialism and hint at why I believe it has great potential for determining the most basic secrets of the universe.

# *Section II*

# Vecton Theory

## A Sample Reality Model

# The Cosmology

### (Step One, Phase One)

### THE VOID

The vecton theory presented here traces what could be the path of a natural physical evolution. If this evolution is portrayed as our cone of categories of Figure 2, then the tip of that cone would have to be complete nothingness, a void of empty space. That the universe began from a void is not self-evident. However, because it seems logical to many that all of the universe should be complete nothingness, the fact that it is not has made popular the belief that physical existence was preceded by universal void. The justification for this tenet turns on the opinion that suggestions of an eternally past physical reality beg the question of primary cause. We will face the question of origin by premising a primordial void.

It might at first seem a little ridiculous to attempt a plausible explanation for the emergence of a complex universe evolving from a void. However, the task is not really impossible if we can establish that void has at least one property as its natural expression or essence.

This property would have to be so simple that it cannot be defined in terms of anything simpler. We are going to suggest that this property is universal emptiness.

**Postulate Number One**

*The universe begins as a void of universal nothingness.*

Universal nothingness certainly isn't much to work with, but there are some things that can be said about our void. It can be said that it has the measure of infinity. We might say that it has a net value of zero and that it must always have a net value of zero. It can even be suggested that the entire concept of universal nothingness is self-contradictory. How can something be everywhere and yet be nothing? How can it be infinite and zero at the same time? What if this self-contradiction was, in reality, meaningful and not just one of human semantics. We will suggest that universal nothingness exists as a zero of two equal opposites, both of which express infinity.

The intrinsic nature of such a void would have to express itself as law. Certainly this void would not allow the sudden appearance of elephants or houses. It must have some form of intrinsic rule which guarantees both its emptiness and its infinity. Since such a rule would have to define two separate, yet mutually expressed, aspects of this one void, it would have to exist in two parts: one part which defines the qualitative emptiness of the void in terms of the quantitative infinity, and one which defines the quantitative infinity in terms of the qualitative zero. The innate self-contradiction of universal nothingness, i.e., infinite zero, leaves us defining zero in terms of infinity. Zero alone can't be nothingness; it must exist for the void as the sum of two equal and opposite infinities. Accordingly:

**Postulate Number Two**

*The universality and nothingness of the void are mutually expressed as the achievement of law.*

**Postulate Number Three**

*The first property of law is the law of zero which states that every quantitative achievement of reality must exist as the sum of two equal and opposite properties so that the net value of the universe is always conserved as zero.*

### Postulate Number Four

*The second property of law is the law of infinity which states that every qualitative achievement of reality must be universal.*

Lest the words achievement and property cause some confusion, the following definitions should be kept in mind.

### Definition Number One

*Achievement*

*The term for any set of two equal and opposite properties. The classifying term for a defined level or stratum of complexity in a consequentialistic reality continuum.*

### Definition Number Two

*Property*

*Any universal expression, value, entity, characteristic, or essence capable of influencing any other universal expression, value, entity, characteristic, or essence.*

To this point in the model we have

a VOID of
UNIVERSAL NOTHINGNESS
expressed as
LAW,
the law of INFINITY and the law of ZERO.

This theory suggests then that universal nothingness naturally establishes two statements or laws by the very nature of its infinite emptiness. This is to say that, contrary to popular human notion, nothingness cannot exist without definition or law. Lawlessness would mean that anything could happen and the nothingness would begin to express "things". Let us now examine the laws of zero and infinity.

## The Law of Zero

What if all conservation laws in the natural sciences are predicated on the demand of a fundamental void that the net value of the universe always be zero? If all of the participants in any non-universal closed interaction are part of some balanced positive or negative universal entity, then all of those participants must have the same value after the interaction as before. If any unbalanced changes at all take place, then one of the opposites is no longer equal in the scheme of all things and a net value of zero ceases for the universe. Since zero must always be zero and no net value greater than zero is allowed in the universe, all properties are conserved in any interaction. This is why any balanced equation always yields zero when both terms of the equation are placed in summation. All equations are statements of conservation.

According to this theory, the nature of void, as defined here, naturally sets up a kind of Hegelian thesis — antithesis momentum, based on the law of zero as postulated. Every property naturally mandates an equal and opposite property with which it must be mutually expressed in summation. This theory suggests that any property may be created on a universal basis providing that at the same time an equal and opposite universal property is created.

The two equal and opposite properties are together called an achievement. Each achievement actually contains three elements: the property, the antiproperty, and some form of mutual expression which is the summation of these two properties. This mutual expression then becomes a new property of the universe which automatically mandates its own anti-property according to this theory.

This concept of void, if a correct one, means that evolution is not limited only to biology but applies to the inorganic sciences as well. This is to suggest that, just as certain molecules have properties which allow an evolutionary progression to an amazing array of complex life forms, so does void have properties which naturally bring about other properties such as direction, distance, and motion. These quantities would then in turn naturally combine and interact to form particles, atoms, molecules, compounds, and material structures.

We will, from this metaphysical yin-yang, suggest that positive and negative charge, quarks and antiquarks, and particles and antiparticles are all derived from this sequence. How is zero maintained during all of this activity? Consider charge: if there is exactly as much positive charge in the universe as there is negative charge, then the net charge of the universe is zero. This zero is not nothingness, because charge exists. However, when the universe is considered in its entirety, there is no resultant charge. To understand this is to understand how this model interprets zero.

### The Law of Infinity

The human mind has always had difficulty dealing with the concept of infinity. This writer takes the position that some of the difficulty with the concept of infinity stems from the fact that it has been misdefined and misused. This theory brings certain clarifications and simplifications to the definition of infinity which might make it a little less mysterious. The question we must ask ourselves here is, not how humans define infinity, but: How does reality define infinity?

The first distinction we must make is between "infinity" and "the infinite". Infinity refers to a numerical quantity which is unbounded. Infinity might come about through some form of repetition, such as that expressed by the decimal form of two-thirds; or it might evolve from the characteristics of some infinite form; for example, the number of meters in an infinite straight line equals infinity.

The infinite is an expression of unlimited size. The universe has infinite size and any straight line can extend an infinite distance. There is no mention of number with the infinite. Very few things are really regarded as infinite; empty space, the distance of an unlimited straight line, and time are about the sum of the list.

This theory takes the position that both infinite distance and infinite time are derived from the universality of the primordial void. On this basis we suggest that there is really only one kind of infinite dimension. What we mean by this will become clearer as the theory unfolds.

This theory is going to postulate that infinities may be relative to each other. This view is mathematically original but cannot be contradicted by anything in our observed universe.

**Postulate Number Five**

*Numerical infinities exist as relative quantities.*

Consider again a finite point A radiating an infinite number of straight lines in all directions. Now consider an identical point B, not A, also radiating an infinite number of straight lines. In classical mathematics the sum of all straight lines in points A and B combined would equal the number of lines radiating from only one of the points. Infinity once is equal to infinity twice no matter what, according to most mathematicians, since multiples or fractions of infinity cannot be measured any more than a single given infinity. We do not believe this view of infinity is logical or in keeping with reality. We will suggest that it is meaningful to say that point A contains half as many straight lines as points A and B combined.

As we have already suggested, there are two kinds of infinities in reality, the infinity of unlimited size and the infinity of unlimited count. However, the definition, unlimited, is a poor one for numerical infinities because of postulate five. For example, it is impossible for point A to radiate more than half the number of straight lines represented by points A and B combined. Hence, in reality, any given infinity is always bound by that which is most possible. Accordingly:

**Definition Number Three**

*Infinity*

*That which is most possible.*

From this, the definition of the finite naturally follows:

**Definition Number Four**

*The Finite*

*That which is least possible, or any real number multiple of that which is least possible.*

There is nothing in any observed aspect of the universe that prohibits this interpretation of infinity and the finite.

As already suggested in our brief discussion on the law of zero, the universality of void and the nothingness of void bring about a third term which is their statement of mutual summation. Consider the following:

> For every property in the universe there must be an equal and opposite property.
>
> +1 yields −1
>
> These two properties must be mutually expressed so that the net sum of the universe is equal to zero.
>
> +1 "plus" −1 yields 0

The third term we are referring to is the "plus" in the above formula. We suggest that the nature of the "plus" changes for each achievement in that each plus is indeed a real relationship and not simply a mathematical term. This will become clearer.

### Definition Number Five

*Mutual Expression*

*The method of summation used by any two equal and opposite properties to achieve the net value of zero.*

This leads us to ask: In what form are the law of zero and the law of infinity mutually expressed? If such a mutual expression does occur, then it would have to give rise to a third kind of thing, a third achievement of the void. This achievement must exist as two equal and opposite properties, both of which adhere to the two laws of void. We will postulate this new achievement to be space.

### Postulate Number Six

*The law of zero and the law of infinity are mutually expressed as space.*

### Postulate Number Seven

*Space exists as the sum of two universally equal, but opposite properties which are unconfined volume and confined volume.*

Here, we are departing from the concept of volume as the mathematical product of depth and area. We are using the word more in its popular sense as expressing available room or space. This concept is probably not as difficult to grasp as that of how such space can express both unconfined volume and confined volume simultaneously. To deal with this, let's put them together in mutual expression. The only physical thing that comes to mind which expresses both a confined volume and an unconfined volume is a soda straw or unending tube. Its unending direction would represent the unconfined volume and the diameter of the straw or tube would establish the spatial definition necessary for confined volume. This means that direction would have to be the natural consequence of space which is both confined and unconfined. Hence, space is not infinitely divisible, but instead exists in straw-like units which are quantum in nature.

It is not difficult for the human mind to abstract a point or a line as being volumeless. However, what we must deal with here are not the capabilities of the human mind but the actual methods of reality. If a dictionary says direction is expressed without volume, does this mean this is the case in reality? Of course not. Can you physically portray direction without utilizing something having volume? Even the ink in your pen has volume. Think about it . . . it cannot be physically done. Therefore, if direction is a physical reality and not just a conceptualization, it must express volume.

### Postulate Number Eight

*The equal and opposite properties of unconfined volume and confined volume are mutually expressed as the achievement of direction.*

### Postulate Number Nine

*Direction exists as the sum of two universal, equal, and opposite sets of directions, one set being positive and the other set being negative.*

### The Corollary of Postulate Number Nine

*For every direction in the universe there exists an equal and opposite direction.*

At this point we have a universe of nothing but empty space in every possible direction. Its measure is infinite and its net quantity is zero. Nine postulates and we still have nothing more than one of the popular definitions of a primordial void!

Our law of zero requires that every direction must be mutually expressed with its equal and opposite direction. This mutual expression is a straight line. A straight line is the simplest possible linear dimension in mathematics. We will refer to it then as a dimension. The only departure we make from its classical interpretation is that we suggest every straight line must express volume since we are forming these lines from directions which do.

### Postulate Number Ten

*Every positive direction is mutually expressed with an equal and opposite negative direction as a dimension.*

If you had a large collection of identical and perfectly straight straws, each having volume, then you could express any area or "volume" simply by aligning or stacking the straws to any form of your choosing. This theory sees the universe as a haystack of such straws, each straw being a unit of dimension. This means that we have a one-dimensional universe with each dimension expressing space (volume). In reality, two dimensional and three dimensional forms may only exist as geometric multiples of this one dimension. It's like buying pipe by the lineal feet and then stacking them side by side to form area and on top of each other to form "volume".

This theory has a completely different view of volume from that of the traditional. We see volume as an intrinsic part of direction and distance and incapable of being defined by anything simpler except law. Classical thought sees volume as a manifestation of relationships between areas, curves, straight lines, and the like. This writer believes this view of volume is based upon our perception of everyday objects and is not valid at the most fundamental levels.

In discussing the notion that the mutual expression of opposite direction forms a straight line, it must be kept in mind that this model gives us two kinds of directions, one positive and one negative. Both kinds must occur in infinite number. Therefore opposite direction cannot be viewed as just another positive direction; it must be held to be truly negative. This then raises the question of how an unlimited straight line can be held accountable to the law of zero. It must exist as two equal opposites and yet the oppositeness of direction is not available since it was used to achieve oppositeness for direction.

One way out of this dilemma is to treat an unlimited straight line as only one property of a dimension, suggesting that there exists for every unlimited straight line an equal and opposite limited straight line. That is, for every infinite distance there is a finite distance.

### Postulate Number Eleven

*Dimension adheres to the law of zero by requiring each individual dimension to exist as the sum of an unlimited straight line and a limited straight line expressing infinite distance.*

### Postulate Number Twelve

*Dimension adheres to the law of infinity by requiring that the number of dimensions in the universe equal half the number of all possible directions in the universe.*

### Postulate Number Thirteen

*Every unlimited straight line exists as an infinite dimension which expresses the greatest possible volume that can possibly be expressed by two and only two opposite directions.*

### Postulate Number Fourteen

*Every limited straight line exists as a finite dimension which expresses the least possible volume necessary to express two and only two opposite directions.*

The progression of our postulates has brought us to another seeming contradiction. We are dealing with a finite distance which the law of zero says must express itself over an infinite distance. Before we address this paradox, it might be helpful if we try to determine the nature of the finite distance itself.

Our finite dimension can only be defined in terms of the properties we already have. Therefore, we can say that it has volume and exists in infinite number; and its form must be linear since it expresses two equal and opposite directions. Beyond these things, there is very little we can say about finite dimension by way of description. However, one thing is clear: it is distinct from infinite dimension because it is the least possible form capable of expressing two directions, while infinite dimension is the greatest possible form in which only two directions can be expressed.

So we have two opposite kinds of dimensions. Both have volume, both have infinite number, both express two and only two directions, but one is infinite and one is finite; and so they are distinguishable, one from the other.

One of the reasons it has been difficult for the human mind to deal with the question of primary creation is man's conception of elementary particles. At the mention of a finite entity in empty space, the mind perceives of some iron-like little bead dangling in suspension. It is difficult to imagine how such a something could come from nothingness. Even if granted that volume, direction, and straight line distance are a part of the universe of void, those things are still a collection of "nothingness," so how could even they form "something"?

According to this theory, inertia, attraction, and repulsion, the three aspects of physics that allow us to perceive mass, arise from interactions between finite entities and not from some form of intrinsic hardness or mass. This notion is not in keeping with our everyday experiences, and our minds keep insisting that matter must ultimately be made up of tangible bead-like entities.

Consider the field of a permanent magnet. It is not tangible in one sense, yet, it does influence certain kinds of matter. No one thinks of it as being made up of little bead-like entities. Yet, in our matterology we will attempt to show how the matter of the proton and electron could be made of this same kind of "stuff".

What we are trying to say is: substance does not have to be substance in the classical sense; it only needs to be distinctive from volume, direction, and straight line dimension, which are just as "real" as substance. There are no tangibles and intangibles; there are only sets of properties.

### Definition Number Six

*Substance*
*The distinction between infinite dimension and finite dimension.*

We recognize the semantical argument can be made that our notion of substance is tantamount to tangibility. We have stressed these distinctions so that the reader will begin to think of substance, not as the classical hard little shot, but as a unit which is just as intangible as it is tangible: as a unit that might be both matter and energy.

There are several things which must be commented upon with regard to each unit of finite dimension. Firstly, their size is absolute. You may imagine them as large or as small as you wish. With regard to their size, all we can say is; there is no smaller dimension capable of yielding two directions in alignment.

Secondly, the law of zero requires all opposite properties to be equal. How can something finite be equal to something infinite? Somehow each finite dimension is going to have to express itself with the entire length of the infinite dimension with which it is associated. This means that finite dimension alone cannot equal infinite dimension. It is going to have to have some "multiplier" or "license" which will let it, a finite distance, cover as much distance as an unlimited straight line (infinite distance). In other words, finite dimension needs infinite time to do what an infinite dimension can do without time. Time for the finite dimension is an obligation to cover all of the distance covered by a unit of infinite dimension.

Hence, a finite dimension is discrete as substance not only by the fact that it cannot be the same thing as an unlimited straight line of space, but also by the fact that it has time. Time, then, must exist only for finite dimension.

**Postulate Number Fifteen**

*Each unit of finite dimension is made equal to its corresponding unit of infinite dimension by means of an obligation on the part of finite dimension to cover an infinite distance.*

This idea that time exists as an obligation of finite dimension to express itself over an infinite straight line will become clearer as the theory unfolds. Before getting too deeply into the model, it might be helpful to outline the achievements and properties of reality to this point.

We begin with
a VOID of
UNIVERSAL NOTHINGNESS
mutually expressing
LAW as
the law of INFINITY and the law of ZERO
which are in turn mutually expressed as
SPACE
which is
UNCONFINED VOLUME and CONFINED VOLUME
which are in turn mutually expressed as
DIRECTION

which is

POSITIVE DIRECTION and NEGATIVE DIRECTION

which are in turn mutually expressed as

DIMENSION

which is

INFINITE DISTANCE and FINITE DISTANCE

multiplied by

TIME

which are in turn mutually expressed as

______________

______________ ______________

The only plausible way finite distance can be mutually expressed with infinite distance, in such a way that they are both made equal through the obligation of the finite to cover an infinite distance, is for motion to occur. Motion, then, must be the form of the mutual expression of infinite distance and a finite distance obligated to express infinite distance.

**Postulate Number Sixteen**

*Infinite dimension is mutually expressed with finite dimension and its obligation to express infinite distance as motion.*

This establishes absolute motion as a new achievement for the universe. This means that for every motion in the universe there must exist an equal and opposite motion.

**Postulate Number Seventeen**

*For every absolute motion of finite dimension there exists an equal and opposite absolute motion.*

It appears that the logical progression of our theory again has us in a little trouble at this point. We have one finite dimension required to express two opposite motions. Our concept of forward and reverse is sufficient for two equal and opposite motions. What bothers us is how we are going to get a single unit of dimension to do both at the same time. The only way this seems possible is for the finite dimension to divide in half, each half selecting the direction it represented in the alignment of the finite dimension. (See Fig. 4.) This division of the finite dimension would not be a physical splitting but simply a natural disassociation. The finite dimension was already divided at its forming in that it had to express two directions, one positive and one negative.

**The Concept of Dimension**

The infinite straight line dimension

The finite dimension

Figure 4.

This means that motion as the mutual expression of finite and infinite dimension is a more complex thing than just forward and reverse. Certainly being half a finite dimension has to be different from being a complete finite dimension. Even though both halves of the finite dimension have the same absolute speed, they express opposite direction and opposite motion.

Our minds want to tell us that opposite motion is fulfilled by opposite direction. But our theory has these as entirely different properties. We have no way of expressing opposite speed. Therefore, we are left to conclude that opposite motion actually imparts a character difference to each finite dimension. This character difference directly results from the positive and negative aspects of the two opposite directions.

**Postulate Number Eighteen**

*Equal motion is maintained by an absolute speed predicated upon the relationship between finite dimension, its obligation to cover an infinite distance, and infinite dimension. This same relationship exists between half a finite dimension and the direction with which it is associated.*

**Postulate Number Nineteen**

*Opposite motion is achieved by the character distinction created by the division, to two opposite directions, of a single finite dimension. Every finite dimension is the sum of two exactly opposite characters, each distinctive from each other and from infinite dimension.*

At this point in the development of our theory we have two kinds of oppositely charactered finite half-dimensions. One set is associated with positive direction and one set with negative direction. Both sets are equal in number and both sets travel at the same absolute speed. Each unit is associated with one and only one direction, and each unit has a volume equal to the least possible volume necessary to define half a dimension. Because their length may only imply one direction, they must be regarded as point-like entities. We will call these point-like entities "vectons".

## THE VECTON

**Definition Number Seven**

*Positive Vecton*

*A point-like entity, being half a finite dimension, expressing: a least possible measure of volume, a single positive direction, finiteness, a single absolute speed, and a positive character intrinsically related to its association with positive direction and the nature of motion.*

**Definition Number Eight**

*Negative Vecton*

*A point-like entity, being half a finite dimension, expressing: a least possible measure of volume, a single negative direction, finiteness, a single absolute speed, and a negative character intrinsically related to its association with negative direction and the nature of opposite motion.*

**Definition Number Nine**

*Diston*

*The amount of distance — volume expressed by one vecton. Absolute Distance. A quantum unit of finite space or volume.*

The two equal and opposite properties of motion are complex in that what makes them equal, speed, is not what makes them opposite, character. Hence, motion is both speed and character. Such dualities might exist for all properties, but their intrinsic simplicities do not let us see them. For example, we might say directions are equal by virtue of their "point" and opposite by their alignment. This much is certain: the concept that positive and negative character is just as much a part of motion, as is speed, will seem strange to most upon first hearing it.

**Definition Number Ten**

*Character*

*A reference term denoting the positiveness or negativeness that arises from the two opposite motions of a finite dimension and positive and negative direction.*

Even though motion is both character and speed, we will generally use the word motion as referring only to speed unless otherwise indicated.

Our theory needs a linking postulate at this point. If we are going to relate the ideas here to our observed world, we need to point to something in our theory that is like something already observed in the universe. We can't prove such a linkage. Its truth must depend, like the rest of the theory, on quantitative investigation.

**Postulate Number Twenty**

*The Linking Postulate*

*The single absolute speed of the vecton is the theoretical speed of light.*

This theory meshes with the theory of special relativity in several interesting ways. Like special relativity, we see mass and energy comprised of the same stuff, and too, we have one absolute speed for energy which is not predicated upon the motion of any other body, only itself. Matter and energy will be dealt with later. Let us now look at this thing of time and relativity.

## VECTON RELATIVITY

At first thought, with its infinite past and infinite future, time would seem a likely candidate to be a category. However, this theory takes the position that absolute time is indistinguishable from the absolute speed of the vecton.

As will become clearer when velocity is discussed, we see, in this theory, only one vecton speed. Vecton velocity then becomes relative speed. Since absolute speed and absolute time are one and the same, velocity also yields relative time.

**Definition Number Eleven**

*Time*

*The measure of the obligation of a finite unit of distance to express itself over an infinite distance. The measure of absolute speed.*

At this point we have two oppositely charactered vectons expressing time-motion as an absolute speed. They are moving away from each other (Figure 4) in exactly opposite directions. We must now ask: How are these two equal and opposite motions mutually expressed? Our answer is that they are mutually expressed as velocity.

We are going to depart slightly from the classical definition of velocity. We will suggest that a single vecton cannot alone express velocity even though it has an assigned direction. We will take the position that velocity can only be expressed with two or more traveling bodies and is a vector statement denoting the degree of change in their relative positions. Accordingly:

**Definition Number Twelve**

*Velocity*

*The measure of change, in terms of speed, between two or more finite bodies induced by the factors of direction and motion. Relative speed-time.*

**Postulate Number Twenty-One**

*Opposite motions are mutually expressed as relative velocities.*

**Postulate Number Twenty-Two**

*For every positive velocity there is an equal and opposite negative velocity.*

To a single free vecton there is no such thing as relative speed. There is only one speed which is the transitional motion of absolute time. However, vectons are constantly changing their relative positions due, not only to the fact that they are in constant absolute motion, but as well to the fact that each vecton is associated with a different direction. The relativity arises from the multiplicity of directions and not from the speed alone.

Hence, to the vecton, relative motion is expressed as velocity. Vecton speed is always absolute and relative to nothing.

To get a clearer picture of this, let's imagine two vectons traveling in exactly opposite directions. Their relative speed is 2c (c being the symbol for the speed of light) but, individually, their absolute speed is only c. In this theory we would say they have a positive velocity of 2c.

Now take two vectons traveling exactly head-on. We would say they have a negative velocity of 2c. Finally, we will think of two vectons traveling side by side in exactly common directions. Their velocity is zero. Hence, any velocity may exist between 2c and −2c.

In the theory of special relativity it is impossible for an observer to perceive of any speed faster than the speed of light. If our vectons do, as we suggest, travel at this speed, then our concept of velocity would at first seem to be in conflict with that theory. However, this is not the case.

In special relativity no observer may "see" anything moving at any speed faster than the speed of light, because the "message of vision" carried by the photon cannot itself travel faster than this speed. Our theory will arrive at the same conclusion when we discuss the electromagnetic wave. For now, we are not concerned with the mechanics of wave propagation. A single vecton is not a photon and is not governed by the more complex relationships that govern particles and photons.

Further, special relativity treats relationships from the point of view of an observer. This writer questions the validity of that approach at the most fundamental levels. Therefore, it is not contradictory to acknowledge that no observer can visually perceive of any speed faster than the speed of light while still maintaining that the real velocity of two head-on vectons is minus twice the speed of light.

To the free vecton there is only one speed in the universe and that is the speed of light. All other forms of motion must be relative to it. In this theory, direction, not speed, determines the vector magnitude of velocity for the vecton. Velocity is always relative for the free vecton and speed is always absolute. The greatest velocity possible is plus or minus twice the speed of light and the least velocity possible is zero velocity. Any measure of velocity may exist between these values. The property of positive velocity and the property of negative velocity is, together, the category of relativity.

Because velocity is relative speed, and time is the measure of speed, velocity fragments speed and time. By choosing this set of vectons, or that set of vectons, we can establish net values based on net directions and net motion. The vecton, or vecton sets, used in such determinations are called frames of reference. Velocity allows us to compare frames of reference.

A frame of reference, then, is the definement of one vecton or a finite set of vectons. If only one vecton is involved, then the frame speed is always c; the frame size is a diston; the frame direction is the direction of the vecton; and the frame character is the character of the vecton. If the frame involves more than one vecton, then the frame speed, size, direction, and character are expressed as the "average" values of these properties, as held by all of the vectons in the finite set.

Time, as vecton motion, does not occur in finite natural units. It is a continuum whose only natural unit is eternity. Therefore, if time is the natural measure of motion, then we need some artificial system for comparing the net motion (speed-time) of reference frames. In everyday experience we do this with the concept of distance and time. However, our time is the speed of light and not distinguishable from the motions we are dealing with. We can ask: "How much time?" only by asking: "How much motion?" This means that we can only define relative speed-time in terms of distance comparisons.

Consider the pattern of vectons expressed in Figure 5A. Each arrow shows directional path. Each vecton is traveling at the absolute speed of motion. The only way we can define the amount of motion is by the degree it causes the relative positionings of the vectons to change (Figure 5B).

**Vecton Speed-Time**

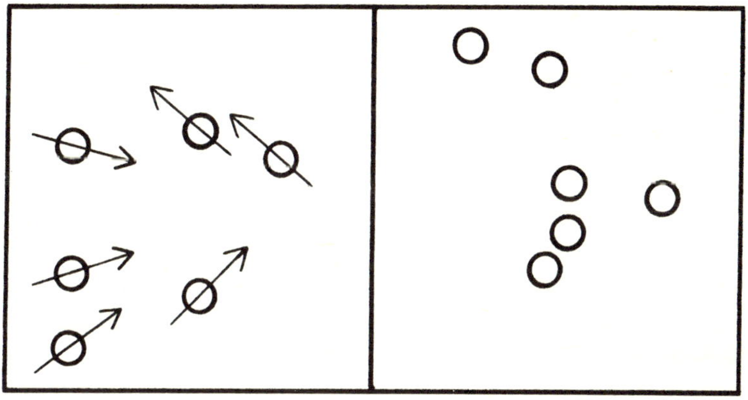

A B

Figure 5. Time is a function of vecton motion and its speed is the absolute speed of that motion.

Just as velocity introduces relative speed-time to the universe, so does it also introduce relative distance. Relative distance establishes pattern.

**Definition Number Thirteen**

*Pattern*
*Any finite expression of relative distances.*

## A Pattern or a Relative Positioning

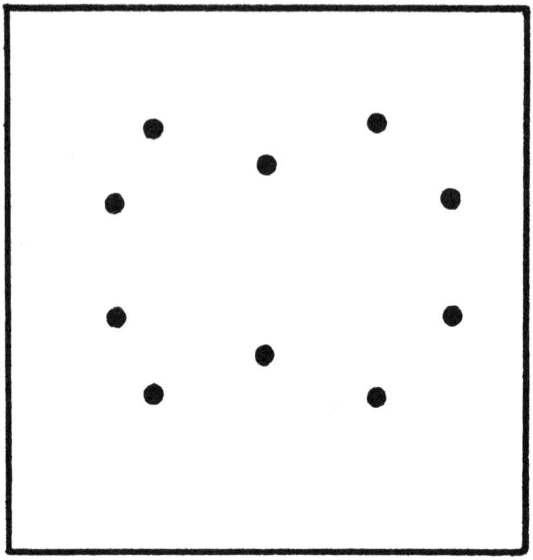

Figure 6.

If the points of Figure 6 are dimensionless, we can only express the distance between these points in relative terms. We cannot tell if the sketch is portraying points billions of light years apart or points only millimeters apart. What we can see is a relative positioning which allows us to recognize a pattern. As long as the relative distances remain unchanged, the pattern will always be recognizable no matter what the absolute distances are.

According to this theory, vectons are not dimensionless but do express an absolute distance based on their quantum volume, which is a diston. Hence, every vecton pattern expresses both absolute distance and relative distance. Now we may ask: Can any pattern of vectons ever truly exist? Because the vectons are constantly in motion, certainly no pattern is ever held except under conditions of common direction. Consider the three vectons in the following sketch.

**A Point of Time**

Figure 7. In this model vectons can express definable relative positions even though in constant motion. However, the concept of "a point of time" is needed.

Vectons A, B, and C are all on the same plane with parallel directions, C being opposite. Will they ever align themselves along a common straight line? Yes, even though the straight line pattern is not held for any duration, it still takes place. To understand this is to understand how this theory interprets five terms: a pattern, a relative positioning, a structure, an event, and a point of time.

### Definition Number Fourteen

*An Event or A Point of Time*

*Any specified pattern formed by any specified set of vectons. Any specified relative positioning of vectons.*

Returning now to Figure 5, the meaning of frames A and B as events should now be clearer. Any time-motion for any set of vectons can be measured simply by comparing any two of the infinite number of events they yield. This measurement is made meaningful since each vecton event expresses both absolute and relative distance.

This kind of reasoning allows us to establish a semi-natural unit of time which we will call a duration. It also allows us to put time-motion limits on any frame of reference in addition to the limits imposed by the number of vectons selected as the set.

**Definition Number Fifteen**

*A Frame of Reference*
*Any finite set of vectons placed within the context of two discrete events.*

**Definition Number Sixteen**

*Duration*
*The measure of all the motion that takes place within any finite set of vectons while a single vecton travels the length of one diston.*

Perhaps Figure 8 might help provide a better feel for the concepts of event, duration, and frame of reference, as well as what these things might imply.

**A Frame of Reference**

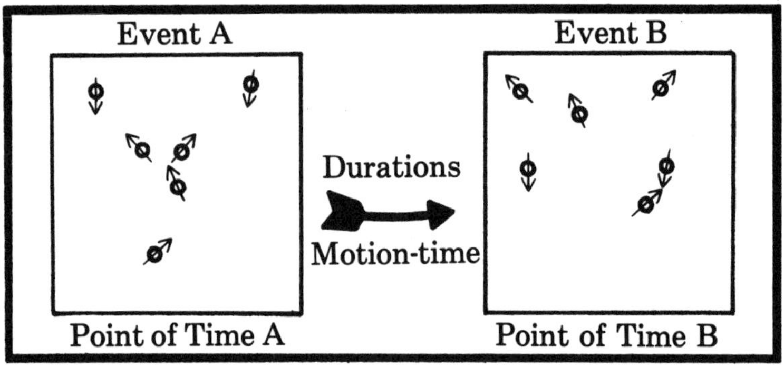


Figure 8. The uniqueness of Vecton Theory forces it to look at concepts in completely different ways. For a specified and finite set of vectons a frame of reference defines not only the vectons and their motion, but all of the patterns or relative positionings between any two specified patterns or events.

## VECTON INTERACTION

With velocity we have opened a Pandora's box of relationships. Vecton groups may have velocities anywhere from zero to plus or minus twice the speed of light; they may express relative time based upon vecton speed; and they may form frames of reference yielding net properties. Truly, velocity is the category of relative things.

Now we must ask: How do equal and opposite velocities mutually express themselves? The only plausible answer seems to be "with vecton collisions." Certainly with all the changes in relative positioning we have alluded to, collisions between vectons must take place.

**Postulate Number Twenty-Three**

*Positive velocity and negative velocity are mutually expressed as vector interactions.*

The fact that vecton groups have net influences based on the summed values of their directions, vecton count, and character, has already been suggested. This suggestion is now affirmed with postulate twenty-three. However, the implications go even deeper because we must conclude that vectons in collision form vector sums.

There are basically two kinds of vecton interactions which may occur: collisions between like-charactered vectons, and collisions between oppositely charactered vectons.

Under the requirements of void conservation the question of what happens when two vectons collide is certainly an interesting one. They must maintain their alignment and their absolute speed. Since they cannot be compressed any further, they must be perfectly rigid.

**Postulate Number Twenty-Four**

*All vectons are perfectly rigid.*

We cannot even plausibly suggest that they pass through each other since that would seem to violate the mutual expression clause of the law of zero and our intuitive concept of substance.

### Postulate Number Twenty-Five

*All vectons are discrete.*

The properties of the vectons correspond with the categories which form them. Therefore, they have volume, direction, distance (speed-time), motion (character), and relativity (position).

The forming of a vector sum must conserve all of these things. Volume would be conserved by rigid discreteness. Direction would be conserved as a spatially defined direction by the inside directions that form it, in a case where the vectons involved are of a like character. Character, of course, would not change and position is conserved as the composite of these properties and speed time. It is easy enough to understand how the forming of vector sum for like-charactered vectons can conserve volume, direction, character, and position, but what about speed-time?

To understand this, let us take two extremes. In the first we will conceptualize two like-charactered vectons tangent and traveling parallel in exactly the same direction. (In the jargon of this model two directional tracks would actually be involved.) It is easy enough to see that their personal absolute speeds would remain unchanged.

Now let us think of two like-charactered vectons having an exact head-on (180°) collision. If we take the position that the absolute vecton speed is not compromised, then we have a problem with direction. The vector direction would be zero. How can speed-time exist if there is no direction to go in? Further, our model has distance and speed-time arising as natural consequences of direction. If direction is compromised, then speed-time must be compromised proportionally. Therefore, if two like-charactered vectons collide head-on, then their direction becomes zero and their speed becomes zero. Here again, zero is the sum of two equal opposites.

From this reasoning we must postulate that the direction of any group of like-charactered vectons having formed a vector sum is equal to the net direction of all of the vectons involved and that the speed of the vecton group is proportional to the theoretical speed of light as the compromised directions of the vectons involved are proportional to common parallel directions. (Popularly: to one common direction.) Hence:

**Postulate Number Twenty-Six**

> *When vectons of like-character interact, their individual properties are compromised as a vector sum which conserves volume and character unchanged and conserves direction and vecton speed proportionally as a binding force and resultant direction and speed.*

To the individual vecton, its relationship to any other vecton or group of vectons means nothing until a collision takes place. Therefore, to a vecton, positive and negative velocity are only meaningful with respect to interaction. Hence, the interaction between two or more like-charactered vectons may in itself be regarded as the property of vector-summation. We will call this property: positive vectoring. (Figure 9.)

**Definition Number Seventeen**

*Positive Vectoring*

*An interaction between like-charactered vectons.*

## Positive Vectoring

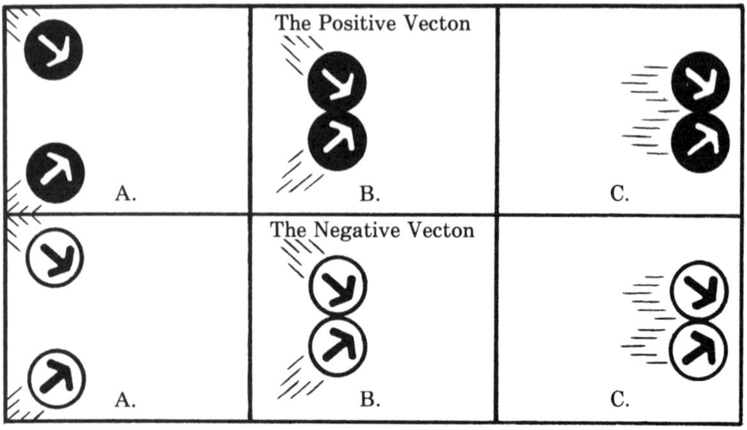


Figure 9. Two like-charactered vectons interact by forming a vector sum which conserves direction as a vector and speed as a binding force.

The compromise direction expressed by the vector sum cannot be a truly new direction for the universe. If the directions of the interactions are to be truly conserved, we will have to suggest that each vecton holds an alignment with "its" direction and that the direction of each vecton is returned to as soon as the vecton group is broken. In other words, in positive vectoring, the vectons individually hold inside direction while jointly expressing spatial direction.

### Postulate Number Twenty-Seven

*Each vecton maintains its directional intent in every case of positive vectoring while conserving its compromised speed as a binding force.*

If positive vectoring is the mutual expression of positive and negative velocity, and therefore a property, then there must exist an equal and opposite kind of interaction called negative vectoring.

Negative vectoring would have to be an interaction between vectons which would yield the opposite of a vector sum. A vector subtraction perhaps? This sounds tricky because the interaction must leave conserved all of the vecton properties involved. This much we can determine, if positive vectoring involves like-charactered vectons, then negative vectoring must involve vectons of opposite character.

Two oppositely charactered vectons upon colliding cannot go off together as a vector sum, nor pass through each other. So what do they do? They exchange that one property which will extricate them from their dilemma. They exchange characters! (Figure 10.)

### Postulate Number Twenty-Eight

*Vectons of opposite character interact by exchanging characters.*

### Definition Number Eighteen

*Negative Vectoring*

*An interaction between oppositely charactered vectons.*

### Negative Vectoring

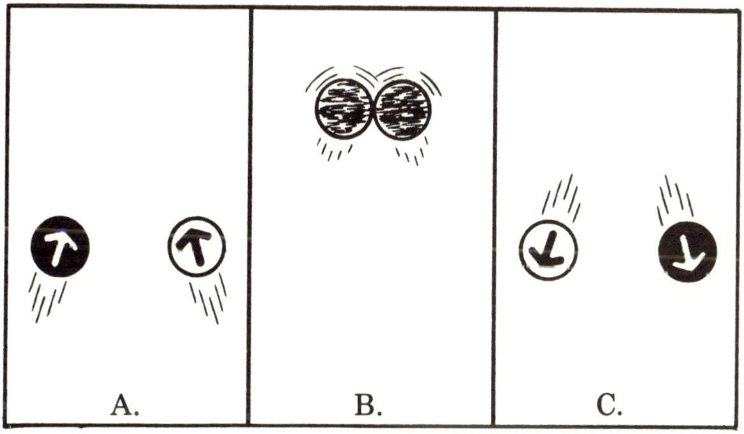


Figure 10. Two oppositely charactered vectons interact by exchanging their characters. Since character is associated with positive and negative direction, the individual vectons must reverse themselves with their new character.

We cannot determine whether the vectons come to a stop during the exchange of character (Figure 10B) or experience no cessation of motion. It would seem that any cessation of motion would defy the law of zero and would be impossible. The law of zero mandates constant motion for finite dimension.

To this point man has only observed in nature two kinds of interaction between discrete units of mass: that of the elastic and inelastic. The elastic interactions, typified by the letter Y, are the interactions of vector summation. These are the interactions of wave reinforcement and positive vectoring. Inelastic interactions, typified by the letters X and K, are the interactions of our everyday experience. According to this theory, X and K interactions are not truly fundamental in that they do not exist at the sub-particle level.

Negative vectoring brings a new interaction to physics, best typified by the letter V. Negative vectoring, according to this theory, takes place only at the sub-particle level. It may be that the success the attractive forces have had in eluding man's understanding is directly related to our lack of awareness of V-type interactions in nature.

In everyday thought, one direction is as good as any other. However, in this theory of opposites, we cannot expect an opposite direction to be just another original direction. This expectation was reinforced at the category of motion when we found opposite motion could not produce opposite speed and relied on the natural disassociation of finite dimension into two opposite characters to achieve oppositeness for motion. Opposite motion could only be expressed with opposite direction.

Therefore, if the values of our postulated universe are to be conserved, we must argue that opposite vecton character automatically implies opposite direction because it is opposite motion. Negative vectoring, therefore, will not conserve the two opposite directions of a given straight line, but it will conserve the exact one for one ratio between the number of original directions and the number of opposite directions.

### Postulate Number Twenty-Nine

*Direction is conserved as long as the number of original directions in the universe equals the number of opposite directions in the universe.*

This postulate serves positive vectoring equally as well. With the vecton traveling in a compromised spatial direction, not the same as its inside direction, it is not truly conserving an absolute direction but an alignment. Therefore, when we speak of vecton character, we are not only referring to positive and negative motion, but to positive and negative direction as well. Yet, the two are so inseparable we may regard them as one thing.

What we have now are two infinitely numbered sets of vectons, one set exhibiting a character opposite from the other. All vectons travel in straight lines at the same speed, each conserving a directional alignment. These vectons are distinct from empty space and have a density in the universe equal to the reciprocal of infinity, which is in reality, even if not in human mathematics, a meaningful relationship. They are perfectly discrete and rigid, having a finite volume which is the least possible expression of volume in the universe. Finally, they interact with each other according to two simple rules of interaction which cause the vectons to either form a vector sum or return along the direction from whence they came.

Now I am going to postulate something that will seem absurd.

### Postulate Number Thirty

*The sixteen properties expressed by the eight achievements of void, law, space, direction, dimension, motion, relativity, and interaction represent the entirety of quantitative properties in the universe.*

If this theory is new to you, there are probably a dozen questions going through your mind right now. The first one may well be: How does this model account for unit charge, quarks, leptons, photons, particle spin, Planck's constant, and all of the other quantities associated with known physics? The second question might be: Why does the property-antiproperty sequence of this theory end with interaction?

Taking this last question first, we are going to reply: It doesn't! Two things should be noted about the sequence of categories as a prelude to understanding the reply. Every category contains the elements of all the categories that preceded it. The laws of infinity and zero, as well as the property of volume, are expressed in direction. Distance expresses direction, the laws, and volume. Motion involves distance, direction, and the laws. In the same sense relativity and interaction involve, either directly or indirectly, all of the properties that precede them. This means that each category is not simply another set of properties but a set of universal complexities which further compromises the laws of the void. Yet, the laws remain in effect, even though the void, as popularly conceptualized, ceased with the property of finite dimension.

If the trend is such that void is continuously redefined by the new properties generated by its law of zero, then its ultimate form would have to be as an expression of all things possible. This expression could only be in one of two forms: either as a single category which allows all things, or as an unending continuum of categories going to infinity. This theory opts for the former.

Therefore, we are going to suggest that positive and negative interaction are mutually expressed as a property capable of yielding all things. It would seem, however, that their mutual expression should be that of a quark or lepton. In a sense it is. But these things are not properties in themselves. We will explain why in the matterology. For now, let us think, not in terms of specific things, but in terms of the logical choice for the property that must follow interaction.

First of all, we will expect this property to naturally express all possible things, or at least be half a category that expresses all things. Secondly, it will have to be comprised of all twelve of the properties that precede it. Finally, the property will have to arise as a natural implication of vecton interaction. The last two categories, relativity and interaction, have not brought new substance but have instead dealt with relationships. The mutual expression of positive and negative vectoring would have to be in this same vein. Therefore, our next property must deal with patterns created by vecton interactions.

We can immediately realize that the number of patterns possible is infinite and that they may express any form that can be expressed by the relative positioning of vectons. In this sense then, the laws of void would be completely open ended and yet still in effect. Further thought lets us realize that probability would be invoked into the universe by such a property. The probability of any particular pattern coming into existence would be inversely proportional to the complexity of the vecton interactions necessary to form it.

**Postulate Number Thirty-One**

*The positive and negative vector interactions of vectons express an array of vecton patterns called physical structures.*

**Postulate Number Thirty-Two**

*The Probability Postulate*

*Any physical structure, not prohibited by the laws and properties involved in vecton interactions, may occur in the universe with a probability of occurrence which is inversely proportional to the complexities of the vecton interactions necessary to define the structure.*

Now perhaps it is clearer why we say that the quantitative properties of the universe end with positive and negative vectoring. All structures that follow are purely qualitative in that they exist by virtue of patterns and not some personal "matter" that is theirs and theirs alone. The matterology will further explain our concepts of quantity and quality.

Here is the balance of our genesis model, as continued from page 65 where we left off with the achievement of dimension.

INFINITE DISTANCE FINITE DISTANCE<br>multiplied by<br>(SPEED) TIME

are in turn mutually expressed as<br>MOTION<br>(THE VECTON)

POSITIVE CHARACTER NEGATIVE CHARACTER<br>(POSITIVE VECTONS) (NEGATIVE VECTONS)

which are in turn mutually expressed as<br>RELATIVITY

POSITIVE VELOCITY NEGATIVE VELOCITY

which are in turn mutually expressed as<br>INTERACTION

POSITIVE VECTORING NEGATIVE VECTORING

which are in turn mutually expressed as<br>PHYSICAL STRUCTURES

ENERGY MATTER

This theory states that every lepton, every quark, every electromagnetic wave, every neutrino, every photon, and everything comprised of these "quantities" are, in fact, patterns formed by the relative positioning of positive and negative vectons. Further, we will state that all four attractive forces in nature arise from the exchange of vectons.

In the matterology, we will now see how vectons form simple particles, complex particles, atoms, and molecules.

# The Matterology

### (Step One, Phase Two)

## IMPLICATION NUMBER ONE
## PARTICLES EXIST AS VECTON CLOUDS

Most, upon initially hearing of my two "particle" cosmology dismiss it off-handedly as being too simplistic and incapable of addressing the complexities of modern physics. However, the implications of vecton theory show at least that a two "particle" universe is plausible. Even if quantitative investigation eventually dismisses our vectons, the truth that many of the complexities of modern physics can be explained with simple laws and structures cannot be dismissed. This theory places too many things, now seemingly unrelated, in logical relationship to allow that.

If the universe is as our theory says it is, then we are in a sense dealing with two kinds of realities. The first reality is that of the void and the vecton. This reality contains the sixteen properties we have already described and may be regarded as the reality of the absolute. These are the things we say are quantitative. The other reality, that of physical structures (i.e., vecton patterns) is qualitative in nature.

The words quantity and quality, as used here, depart somewhat from their classical meanings. However, there are no other suitable terms of classification in the English language and we will have to go with them. Unit charge, for example, has always been thought of as an absolute entity in itself which mysteriously inhabits particles. This theory will later suggest that unit charge arises from the interactions of vectons. Particles and quarks will not be portrayed as little point-like entities but will be viewed by this theory as clouds of interacting vectons.

Analogies are a poor tool in physics but we will try one here. Consider a bricklayer, a large pile of bricks and a vacant lot. The bricklayer knows only two actions, one is to put bricks end to end and the other is to place flat bricks one exactly on top of the other.

Initially, all he can build is a single course of bricks. By making the course the bricklayer has made no new quantities. All he has done is to change the relative position of those bricks used in the first course from the relative positions they had in the brick pile.

By simply repeating this process again and again he now builds a wall. The wall is purely qualitative in that its existence does not add in any way to the quantities involved. For every brick placed in the wall one brick must be taken from the original pile. A building can then be built by repeating these same processes. If the lot is large enough and the bricks plentiful enough, the bricklayer, even though able to perform only two simple functions at one time, can build an entire town of buildings, rooms, walls, walks, and so on.

This theory sees vectons as the bricks, empty space as the vacant lot, and the laws of vecton interaction as the bricklayer. Particles and atoms are then comparable to the courses and walls of this simple analogy. Only vectons have true quantity, all else is of the relative positioning of vectons and is purely qualitative in nature.

In modern theory there are essentially three kinds of basic structures: the quarks, the leptons, and the bosons. There are six kinds of quarks labeled: up, down, strange, charm, bottom, and top. Very recent thought has suggested that each kind of quark might exhibit three "colors." This latter notion we will not deal with here, because colors are not real structures but aspects of the quark structure. There are three kinds of leptons: the electron, the muon, and the tau. The tau is a newly discovered "giant" lepton and is perhaps more a lepton "resonance." Each lepton is associated with its own form of neutral particle called a neutrino. There are several kinds of bosons, each associated with one of the four attractive forces. The photon is a boson.

Each quark and each lepton has an anti-form. Structures formed by quarks are called hadrons. The combination of one quark and one anti-quark forms a hadron group known as mesons. This group includes the pions and the kaons. Three quarks form a particle class known as baryons. Both the proton and neutron are baryons. Three anti-quarks may also combine to form an anti-baryon. No quark has ever been individually observed. The determination that there are six kinds (top is predicted) of quarks is based on a pattern fit with the various types of mesons and baryons. No one can unquestionably state that all six quarks aren't different forms of one or two basic quarks. Figure number 11 outlines the particle groupings.

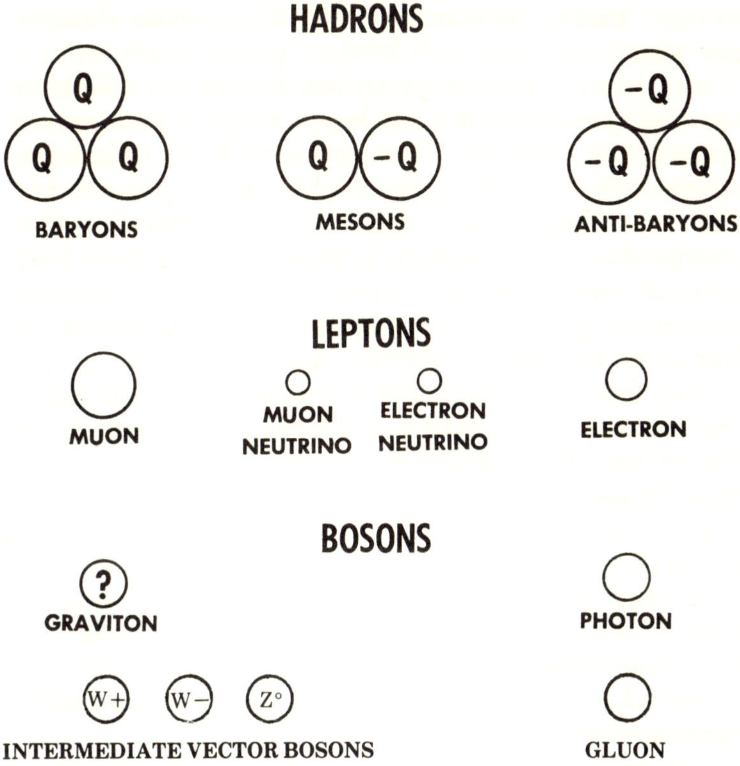


Figure 11.

In electron-positron annihilations, pions are often formed whose typical mode of decay is into a photon and muon. A muon has a lifetime of 2.2 x $10^{-6}$ seconds, and a mass over 206 times that of the electron into which it decays. What happens to the 205 electron masses of the muon?

The decay, as viewed in a cloud chamber, seems completely spontaneous and there are no visible "tracks" of any particles leaving the muon-to-electron transformation. To account for this mysterious disappearance, as well as a similar mass-loss mystery in beta-decays, the concept of a neutrino is used. Using the mass-energy relationship of special relativity, it is theorized that the mass is entirely converted into energy and carried away at the speed of light in the form of a massless, neutral particle.

The primary mystery with neutrino theory relates to the mechanisms of particle decay. Why would a particle like the muon suddenly convert so much of its mass into energy? What mechanism does it use to achieve this? The fact that so much mass can yield so many ergs of energy does not necessarily mean the metamorphosis will take place. Neutrino theory provides no mechanism of particle decay. Modern physics cannot explain why or how a muon forms neutrinos; it can only state that so much mass is lost and offer the forming of neutrinos as the best available explanation. There is not even an explanation for the reason the energy is not all given to the new-born electron.

What neutrino theory has going for it is an impressive set of experiments which show indirectly that neutrinos, or at least neutrino-like entities, do exist. These experiments are a matter of record and will not be dealt with here. Our task at this time is to outline a plausible generalization using our vectons to account for particle structure, particle decay, and the forming of neutrinos.

Our vecton theory provides two kinds of interaction. In one, positive vectoring, like-charactered vectons form vector sums; in the other, negative vectoring, two opposite vectons, upon interacting, return along the path from whence they came. According to our theory, a simple particle, like the electron, is a cloud of billions times billions of vectons held together by negative vectoring.

In this theory a particle is always a cloud of interacting vectons. Our vecton is not a particle; it's not even a structure. At best we might call it a unit of influence which is perceived as mass when interacting in a vecton particle cloud and as energy when traveling freely. To this point we have occasionally used the word particle or structure when referring to a vecton. However, from this point on, particles will always be vecton clouds.

The distinction between particles and vectons must constantly be kept in mind. The laws of quantum theory and special relativity deal at the particle level. Vectons are not particles and can do things particles cannot do. Vecton clouds behave very much as quantum particles do. Vectons themselves do not, however, behave as quantum particles. Vectons can travel at the speed of light; particles cannot. Vecton properties are conserved under the rules of positive and negative vectoring; particle properties are conserved under the rules of quantum mechanics or classical physics. Particles have charge; vectons have character. Vectons are not particles, but vecton clouds are.

To understand how a vecton might form a particle, let's look first at an example of a "closed" vecton interaction. The parallelogram of Figure 12A portrays four equal distances, each the path of a vecton. Remember the character of a vecton is reversed after an incidence of negative vectoring (Figure 12B).

**Closure Under Negative Vectoring**

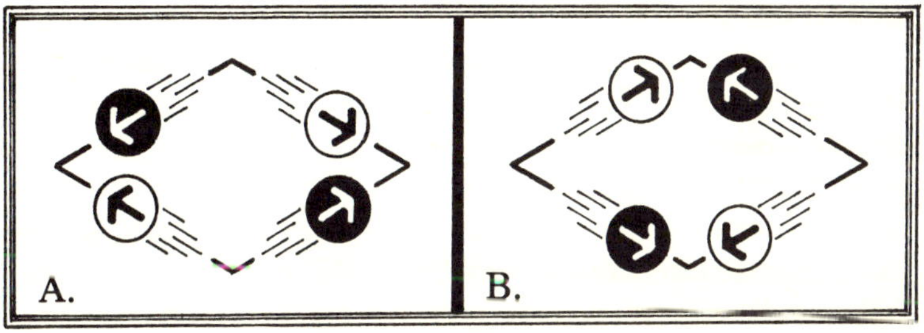


Figure 12. If vectons, alternating by character, form a circular pattern such that vecton escape is limited, individual vectons will move back and forth while vecton character flows oppositely in full circle.

We could call this four vecton set a particle. In reality its occurrence alone would be quite rare and the pattern would probably be disrupted by outside vecton intrusion. As a particle it is going nowhere and its net character is zero. Nevertheless, it does show that negative vectoring can "confine" vectons to a particular region.

Just as the primary law in the reality of the void is the law of zero, the law of probability is the fundamental law in the reality of physical structures. Within limits, the law of probability says: "If you want to make something happen, just make the numbers big enough." Therefore, if I want to create a closed system of vectons, I would have to assemble a collection of vectons large enough that any given vecton would, upon leaving the collection, strike an oppositely charactered vecton, also leaving. The vectons, upon interacting, would exchange characters, and both would return toward the center of the collection.

Actually, it isn't all this simple. Because of the mathematical relationship between a sphere's surface area and volume, it can be shown that any number of vectons, with perfectly random directions, would always have an "escape factor" and the vecton cloud would eventually evaporate.

However, with the vectons there is another set of factors that enters the picture. Each vecton has an absolute volume, hence, any defined vecton cloud always has a surface area which is some real number multiple of the area expressed by the "front" of the vecton. This means that the relationship of the vecton to the cloud size is always finite.

Let us further suggest that our collection of vectons has a directional bias: that is, it originated from some common vecton source. In this case we would not be dealing with all possible "points" at the surface of the cloud sphere, but with only a part of that surface. The two factors of finite ratio and directional bias would greatly extend the lifetime of a vecton cloud in that it would diminish the chance any given vecton would have of leaving the group without encountering an opposite vecton.

Taking the three factors: negative vectoring probability, finite size relationship, and directional bias, we have the foundation for a plausible explanation for particle formation, particle lifetime, and particle decay.

Let's begin with an electron-positron acceleration. Our theory would portray an electron as a huge cloud of both positive and negative vectons with the negative character primarily sharing the direction of the particle and the positron as a vecton cloud for which the reverse is true. From interactions that will be discussed under the electromagnetic force, the electron in a particle accelerator migrates forward by receiving positive vectons in the front and is pushed forward by the positive vectoring of negative vectons arriving in the rear. The reverse would be true for the positron.

Electron-positron annihilation, then, would essentially be a collision between two clouds of vectons. Each cloud would contain billions times billions of vectons of both characters. Each vecton would be a unit of influence.

At impact, billions of individual vectons leave the point of impact at the speed of light. Some form like-charactered vecton clusters of perhaps millions of vectons, but many leave as individuals. Since an electron might contain 2 x $10^{50}$, or upwards, of these vectons, even a cluster of a billion vectons might not be perceived by even the most sensitive instruments.

The vectons, from the collision of the accelerated electron and positron, form a single vecton cloud called a virtual photon. However, both of the original clouds retain enough individual identity that the misnamed virtual photon splits again into two clouds. We will suggest one cloud is a positive pion and the other a negative pion. Let's follow the negative pion.

The negative pion is a seething cloud of vectons. Inside this cloud individual vectons are constantly joining clumps of vectons of their own kind, only an instant later to be touched by oppositely charactered vectons and sent to another part of the cloud. Meanwhile, the surface of the cloud is losing billions of vectons.

Relying on sufficient vecton number, the law of probability, the finite relationship between vecton and cloud size, and directional bias, I am going to suggest that the core of the cloud begins to establish a probabilistic closed vecton pattern, not one as simple as the one in Figure 12, but one comprised of billions times billions of vectons. The closed pattern emerges through a kind of natural selection process for each vecton. By forming rows of alternating oppositely-charactered vectons, the cloud keeps the vectons in these rows from having any chance of escape. Those vectons not finding a seat in the game of musical chairs bounce around the particle at the speed of light until they fall into such a closed niche. If they fall into no closed pattern at all, they will be "selected out" of the particle.

Essentially, closure would be created by a circular pattern of vectons of both characters. If the vecton cloud was comprised of only one vecton character, it would form a uni-directional "clump" whose speed and direction would depend on the net direction of all vectons involved. In a vecton cloud with vectons of both characters we may imagine that both positive and negative vectoring takes place.

In negative vectoring a vecton is always returned along the path from whence it came. If a certain probability of negative vectoring exists, and if the vectons are not free to travel any distance before being negatively vectored, then free vectons can become locked into probabilistic circles. I will call these circles "force lines."

Later we will suggest that such circles form the force lines of magnetic fields. In reality, the body of a force line is perhaps millions of distons across and holds its pattern by the high probability it offers each vecton of negative vectoring. Figure 13B is a slightly better portrayal of a force line than 13A. According to this theory, the holding of vectons in circular force lines through probable negative vectoring underlies the nature of the weak attractive force, while the force lines operating as a whole provide the foundation for the strong force. This will be dealt with later.

**Closure**

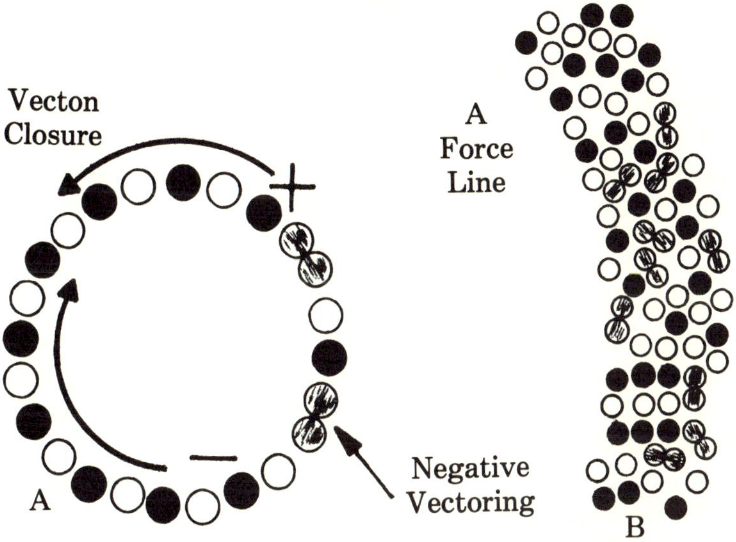


Figure 13. If enough vectons, evenly mixed, are brought together in circular fashion, they will establish an improbability of escape for each vecton. In this theory, such vecton circles are called force lines.

Without quantitative determination, it is only a matter of conjecture as to whether these force lines completely "lock-up" allowing no individual intra-particle vecton movement, or if they are loose associations of vectons traveling very short distances back and forth. Personally, I would favor the "loose association" concept as the more probable.

Thus, in our pion particle it would seem that such force lines would have a certain probability of leakage, at least in the outer regions of the vecton cloud. We may imagine that particles are formed by many force lines. The inner force lines would be much more stable in that they would have more interactions, "being at the center of things."

Following a certain probable lifetime, the pion must reach a point where those vectons in the outer force lines cannot be sustained by negative vectoring. The vecton count has, for them, been so reduced by random vecton loss and the closed vecton patterns at the center of the cloud that the remaining vectons are freed to leave the cloud at individual speed. At this point the cloud disintegrates into essentially two vecton groupings: the set of vectons interacting in the central force lines and the free vectons. These free vectons would still have the directional bias of the pion and would constitute the muon antineutrino.

**Muon Neutrino Formation**

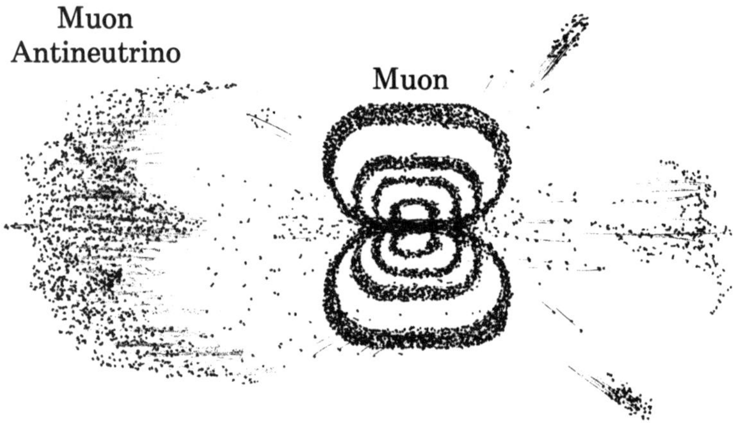


Figure 14. In vecton theory neutrinos are loosely associated groups of like-charactered vectons having a common origin and common direction.

Later we will discuss how a closed system of vectons can be "steered" via negative vectoring toward an oppositely charactered vecton source. Our vecton neutrino has no such steering abilities and is no longer a true particle in that its vectons are not interacting.

Hence, it would not interact with the matter around it in the same way particles do.

It is not likely that our vecton neutrino would have the same kind of lifetime now attributed to neutrinos. It would seem our neutrino would be dismantled vecton by vecton through negative vectoring. However, this would take considerable time, and without steering ability, our vecton neutrino should have much of the penetration power attributed to neutrinos in general. Vecton theory, then, would define a neutrino as a concentrated group of free vectons with a common direction.

In quantum theory, particles spin like tops. Such spin has never been directly observed and exists as a mathematical construct used to balance other constructs and observations of quantum theory. In vecton theory, spin would exist also but not perpendicularly to particle direction. In our vecton force lines we would see character transferred through negative vectoring in exactly opposite directions. The vectons would carry one character when traveling in one direction and the opposite character on the return trip.

In our model, spin would not be as a top spins but perhaps comparable to the spin of a bulldozer tread. Further, the particle itself would not spin, only the character (charge) of the particle would spin. Individual vectons would simply move back and forth.

It is obvious that our model must rely on vecton character as the source of particle charge. What is not obvious is how a particle, equally comprised of both kinds of vectons, may have charge. There are two ways of viewing our model's single cloud particle. One is from the point of view of a given vecton. Here, the vecton vibrates back and forth having one character while in one direction and the other while in the opposite direction. For them there is no spin and no net charge.

The other view is from that of vecton character. From their perspective they flow in circles, trading places with each other at each instance of negative vectoring. The axis along which all force lines join form a zone of strong positive and negative vectoring and enriched destiny for both vecton character and the vectons themselves.

**Closure Character**

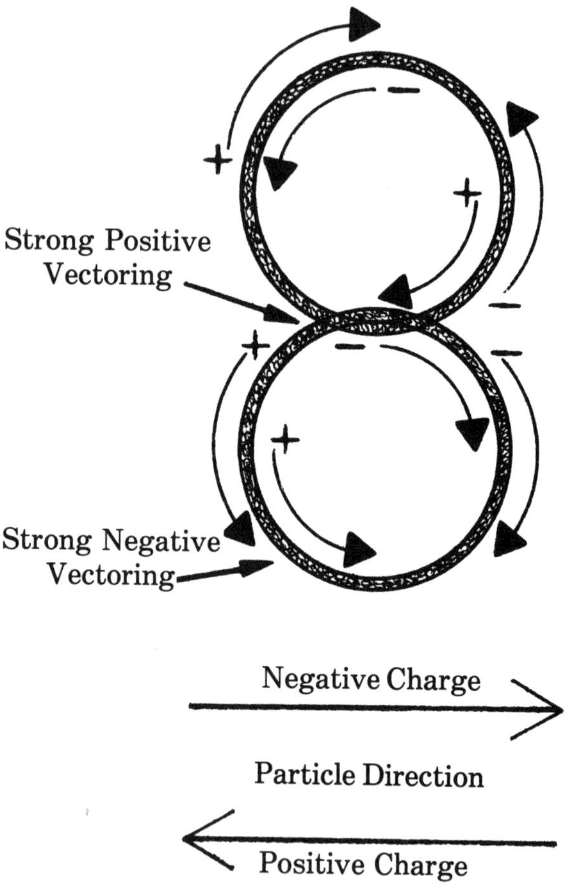


Figure 15. If two vecton force lines intersect as portrayed here, then they will form a two opposite directional character biases.

This model implies that the main body of a particle is its axis. The direction of any particle bound by such negative vectoring would have to exist as a probabilistic composite of all of the vectons in the particle. If a particular general direction is held by more vectons than the others, the vectons having that directional plurality will be able to move in that direction just a little bit further before being negatively vectored. They will thus affect the mean density of the force line, hence transferring and averaging that particular directional bias. Particle speed then is purely a function of vecton parallelism. If the vectons of a particle exhibit all directions evenly, then the particle is neutral and has no direction. If there is a bias, then the particle would move in the direction of that bias.

Because the particle axis defines the particle body, it follows that the axis must always be aligned with the particle's direction. Hence, for any particle in motion it is either moving in the same direction as its positive character or as its negative character. According to this theory any antiparticle is simply a particle put in reverse.

If the central axis of a particle exhibits more positive vectoring than the rest of the particle, it would seem that individual vectons and vecton clumps would, at the end of the axis, have some likelihood of escaping negative vectoring. Hence, our model leads us to suggest that all particles emit unwanted vectons at both ends of their central axis. Positive vectons would exit the positive end and negative vectons would exit the negative end.

The charge of the particle is thus determined by the way the particle interacts with other particles and is steered by its "lead" character. The positive pole, when leading, will be repulsed by positive vectons and attracted by negative, and the negative pole will exhibit the reverse effects.

In simple single cloud particles spin must always reverse when charge is reversed. In particles formed by more than one vecton cloud spin is more complicated. However, our model does not offer right-hand and left-hand spin as defined in quantum mechanics. Quantum unit spin might exist as a mathematical illusion.

Returning to our pion, which now has to be a negative muon if our fit with the observed universe is to be correct, we must suggest that vecton loss continues. It is not likely that its force lines are perfect. However, since the vecton interaction is more patterned for the more compact force lines of the muon cloud than for the pion cloud, the muon can be expected to have a much longer lifetime than a pion.

The concept of force lines as closed vecton patterns is essential to this theory if it is to explain particle structure. Later we will see that a similar notion is needed to explain the nature of a permanent magnet. There are certain likelihoods that might guide any quantitative investigation of vecton closure.

Firstly, a perfectly closed vecton force line would have to involve billions times billions of vectons. The nature of probability mandates this. This is also mandated by any argument that a vecton is a unit of energy. The spiral of an electron toward a magnet does not at all appear jerky. The smoothness of that inward spiral requires a tiny measure of influence if that influence is unitary.

Secondly, this theory must suggest that even a charged particle, like an electron, must contain vectons of both characters. This means that unit charge is an illusion created by the sameness of each vecton, the popular notion of momentum, the popular notion of energy, and the fact that magnets emit vectons. Different numbers of vectons emanating from the magnet are interpreted as magnetic strength. More vectons in the particle are interpreted as more mass, hence more momentum.

The bottom line of all this is that charge is seen as constant for every particle having charge. Magnetic strength, particle mass, and particle energy are all perceived as variables, which they are. However, if all vectons have the same amount of influence; if particle mass is purely vecton count; if a magnetic field is entirely comprised of vectons in motion; if magnetic strength depends on vecton count; and if particle speed is a function of vectons interacting as a group — then all experimental apparatus will be incapable of distinguishing between our vecton interpretation of charge and that of quantum theory.

Thus, we can say that the observation of unit charge does not appear to compromise our vecton theory. Further, if vectons are a reality, then unit charge is an illusion generated ultimately by the unit influence of all vectons. This concept of charge, if true, would explain why the big proton and the little electron both have the same amount of charge influence. It would also help to clarify how a quark appears to have a charge of only one-third. However, it does not seem to readily address itself to the question of why one proton always is associated with only one electron in atoms. This we will take up later.

Returning to our particle decay sequence, we note that the center of our muon is the same as that of the pion. Hence, by this time it has been around for quite a while. As the most long lived and most intense area of vecton interaction, if there is any naturally perfect form of closed probability structure, this zone is our best candidate for finding it. Thus, we must now see three sets of vectons in our negative muon: the emerging "best possible" pattern at the cloud center, the remaining positive vectons, and the remaining negative vectons. (Figure 16.)

**The Vecton View of a Muon**

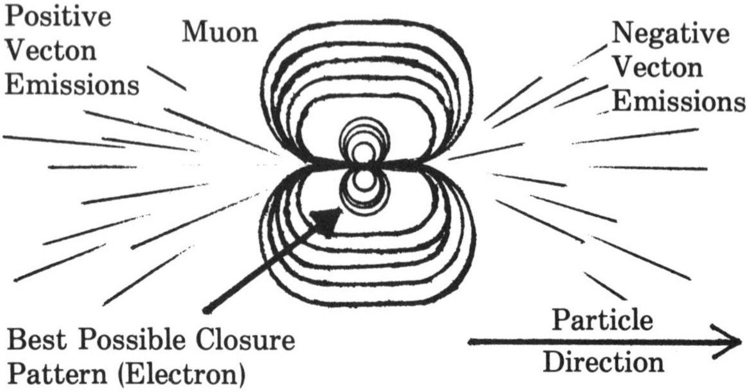


Figure 16. In this theory, the muon is basically an electron with additional unstable force lines which disintegrate into two neutrinos.

Eventually, the same thing is going to happen to our muon that happened to our pion. From its steady loss of vectons, the outer muon force lines will reach a point where they can no longer sustain themselves and will disintegrate. We will expect a muon collapse to yield two neutrinos: one associated with the positive vectons and one associated with the negative vectons. The most perfectly formed vecton force lines would be the remaining structure. This, of course, would have to be an electron.

In the theory of quantum electro-dynamics the electron is portrayed as a point-like object expressing energy without size. So far all experiments have supported this view of the electron. It is a problem for this model because our electron must express volume. It could be that any volume the electron has might be undetectable when viewed from the perspective of electrodynamics. Its own vecton economy might cause it to influence matter much like a photon, thus creating the mathematical illusion that the electron is without size. That is, as the most efficient particle it might express zero size as far as particles go, yet have volume in terms of a sub-particle structure such as a vecton.

Vecton theory has no place to go but to suggest that hadron sub-structures (quarks and/or partons) are lepton-like vecton clouds. When explaining the strong nuclear force, we will attempt to show that the quark binding energy arises from the sharing of force lines. Without the discussion on permanent magnetism we are somewhat disadvantaged here to discuss how this takes place. For the time being we can only suggest that both leptons and quarks, according to vecton theory, will have to behave as tiny permanent magnets.

It would seem that the electron should reach a point of stability such that all of the vectons it contains are guaranteed that they will not be lost at the particle's poles. The observed lifetime of electrons and stable quarks suggests this must be the case. Therefore, we will surmise that electrons only emit vectons when absorbing an excess of vectons.

Figure 17 portrays this model's idea of an idealized "quark" or lepton.

**A Vecton Cloud as a Simple Particle**

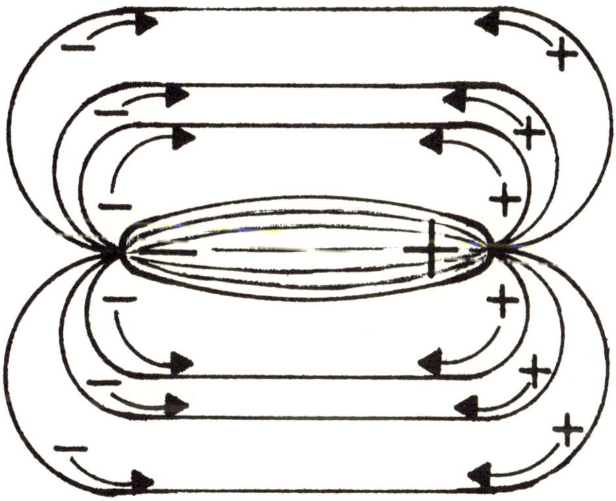


Figure 17.

To the degree that quarks are sub-divisions of complex particles, we are suggesting that quarks are lepton-like structures. In the case of quarks, this spin pattern exchange of vecton character would have to take place from quark to quark. Hence, the vectons are continuously conserved and the quark vecton cloud is prevented from decay. Consider how vecton theory envisions a neutral meson.

**A Neutral Meson**
**(A Complex Particle)**

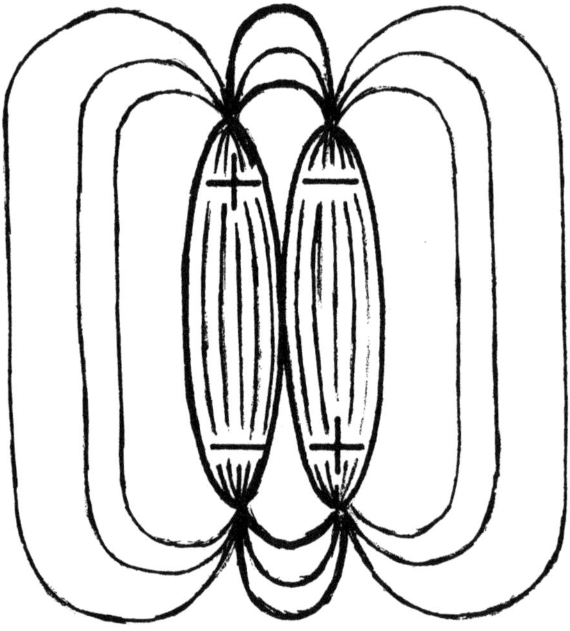


Figure 18.

The sharing of force lines by two or more single cloud particles to form a complex particle might serve to hold the vectons of the outer force lines, thus preventing the disintegration of the quarks into electrons. This brings us to an interesting question: Why doesn't a charged pion decay into two leptons? The answer, according to this theory, is that a charged pion is not a true meson, but a neutral pion is.

Vecton theory departs from quark theory in that it cannot form a one for one ratio between quarks and vecton clouds for all particles. A lambda, for instance, has four vecton clouds but only three quarks. Yet, when the typical modes of decay are looked at, the vecton cloud theory fits perfectly. For example, the typical mode of decay for a lambda is into a proton and a charged pion, and that of a neutral kaon is into two charged pions. Quarks are not conserved when mesons decay into leptons under modern quark theory. The explanations for this are plausible but labored. Vecton cloud theory greatly simplifies the concept of particle decay.

The cloud model serves antiparticles equally well. The concept of quantum spin and unit charge, both which may be illusions, brought to physics the idea that antiparticles are antimatter. This idea is no longer as popular as the realization grows that antiparticles may be nothing more than mirrored particles.

Vecton theory holds that antiparticles are particles in reverse. This may seem like an oversimplification, but it holds even for those neutral particles which are their own antiparticles. Figure 19 portrays our view of some of the more common particles.

## Particles Portrayed as Vecton Clouds

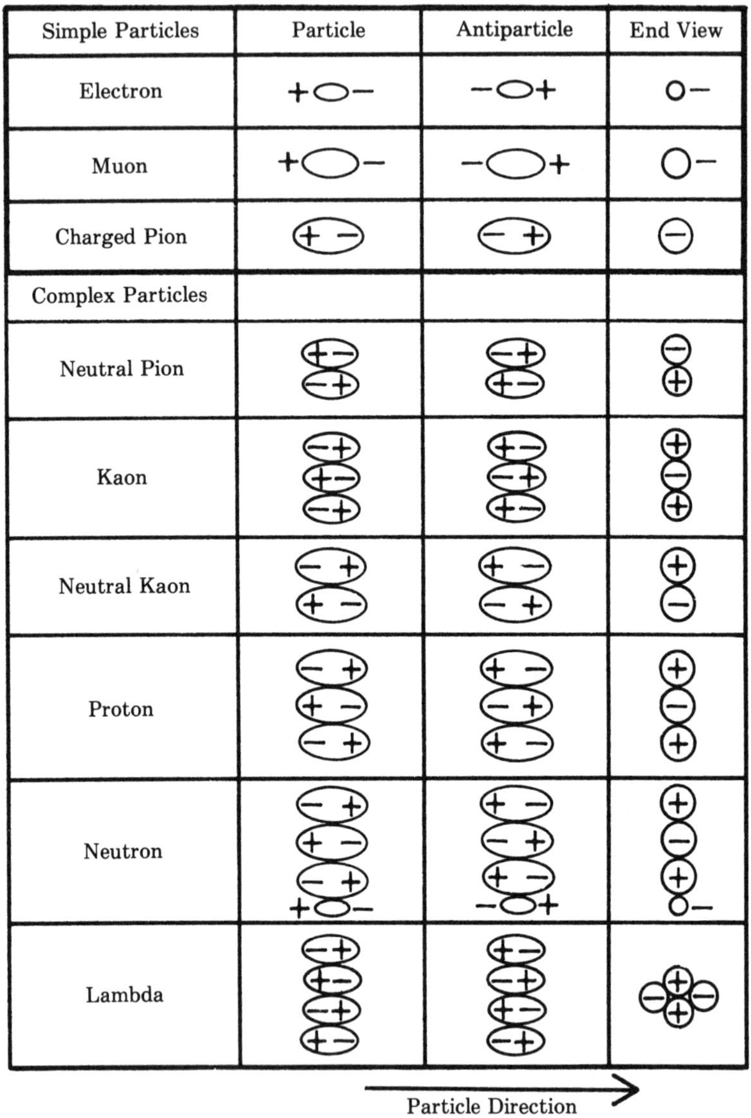


Figure 19.

It might seem that with this particle model, there is no particular reason why all electrons are negative and all protons are positive. However, this is not a serious problem for the theory. If we postulate that galaxies are the eventual products of huge clouds of hydrogen, then we may imagine that such clouds began as a mixture of hydrogen and anti-hydrogen atoms. Opposite atoms, upon colliding, would annihilate each other. The process of majority rule would eventually prevail and only one of the two kinds of atoms would remain. It may be that all of the electrons in some of our neighboring galaxies are, in fact, positrons and the protons, antiprotons.

The builder of any consequentialistic reality model will have to initially make the kinds of superficial qualitative fits with the observations of physics we are making here. It is not easy to deal with such new ways of looking at old concepts. Therefore, you have probably had some problems following my reasoning to this point. Remember, what I am trying to share with you is the kind of thinking I did in developing my postulates and checking them for plausibility. I am not arguing my theory. This step, step one, is for the eyes of the theorist only. Step two, which quantitatively presents the ideas of the model's matterology and offers quantitatively verifiable experiments, must accompany any public presentation of the postulates of the cosmology.

It is thus very important that the theorist begin to take note of any quantitative arguments that might begin to emerge from his qualitative attempts at developing a matterology from the logical implications of the postulates. In this light, let me share with you my interpretations of the success of this particular reality model when I had reached this point in its development. If you, the reader, attempt reality models of your own, such insight may be instructive.

This was the first model I developed from which could be argued a decay sequence which seemed to fit those observed in physics. The fact that it plausibly formed neutrinos was also encouraging. However, it was a great disappointment to me that the model would not yield unit charge for particles, except as an illusion created by the unit character of the vectons, the fact that a particle's mass and momentum are determined by vecton count, and the fact that magnetic force lines are also comprised of vectons.

My concern was not that such an illusion might be impossible but that the community of physicists would offhandedly dismiss any quantitative arguments based on that concept. For this reason I reviewed everything I could about unit charge, even as far back as the work of Millikan. Nowhere could I find any direct observation that could disprove this model's contention that unit charge is anything but an illusion. When the day comes that I want to present this particular model as viable theory, I must be prepared to quantitatively argue the illusion premise.

A second problem with popular thought created by the vecton view of the universe is that the all-neutral particles must consist of an even number of vecton clouds. This magnet view of particles forces the idea that a charged pion cannot contain two quarks if a quark is a vecton cloud. How do I handle this?

The first thing I have to know is that quarks are viewed in two ways. There is the quark as a mathematical construct and there is the quark as a kind of physical bag of matter-energy. Partons are, for example, such theorized sub-particle physical constructs. In the mathematical construct a charged pion has two quarks, but it often decays as if it were only one "bag." I suggest the reader go back to the work of Hideki Yukawa in 1935 and carefully study meson theory in terms of quark theory. Next, study the arguments of theoretical physicists attempting to explain the disappearance of "quarkness" during the decay of a charged pion. I did these things, and I am convinced that physical quarks are only valid when viewed as a kind of bag, vecton cloud if you will.

If you develop a model of your own then, you too, will likely be confronted with similar conflicts between popular theory and your model's interpretation of direct observation. When this occurs, it is better to study your model from the point of view of direct observation rather than popular theory. In the case of my cloud model of particles (Figure 19), I knew that I might prove my hunch about quarks if my vecton clouds succeed where quarks fail, in terms of conservation.

Our model does not let a neutrino be a true particle. Later we will see that a photon is a kind of vecton clump. A virtual photon we will call a concentrated assemblage of vectons which may form one or more vecton clouds. This means that every particle in this theory can be assigned a number based on the number of clouds it contains. Neutrinos have the number zero. Charged pions, muons, and electrons have a number of one; neutral pions and kaons, two; a charged kaon and proton, three; a neutron and a lambda, four; and a charged sigma, five. All particles with even numbers are always neutral; all particles with odd numbers are always charged; and all particle numbers must always be conserved in every typical mode of decay.

The four basic tools of particle decay studies are a cloud chamber, a set of magnets, a clock, and a centimeter stick. Thus, only charged particles can be "seen" and measured. Neutral particles can only be "seen" if they decay in the field of view into charged particles. Photons, unless they decay into charged particles, never leave any trace and therefore can only be surmised. If we take the view that bubble-chamber photons are clouds of mixed vectons, a virtual photon may hold any cloud number depending upon its formation.

With these rules we see most typical modes of decay fit our cloud or bag model of particle structure. Here are some examples:

$$u^- \longrightarrow e^- + \upsilon_u + \overline{\upsilon}_e$$
$$1 \longrightarrow 1 + 0 + 0$$

$$\pi^- \longrightarrow u^- + \overline{\upsilon}_u$$
$$1 \longrightarrow 1 + 0$$

$$\pi^- \longrightarrow e^- + \overline{\upsilon}_e$$
$$1 \longrightarrow 1 + 0$$

$$\pi^0 \longrightarrow \gamma + \gamma$$
$$2 \longrightarrow 1 + 1$$

$$K^+ \longrightarrow \pi^+ + \pi^+ + \pi^-$$
$$3 \longrightarrow 1 + 1 + 1$$

$$K^+ \longrightarrow \pi^+ + \pi^0$$
$$3 \longrightarrow 1 + 2$$

$$K^0 \longrightarrow \pi^+ + \pi^-$$
$$2 \longrightarrow 1 + 1$$

$$\eta \longrightarrow \rho + e^- + \overline{\upsilon}_e$$
$$4 \longrightarrow 3 + 1 + 0$$

$$\Lambda^0 \longrightarrow \rho + \pi^-$$
$$4 \longrightarrow 3 + 1$$

$$\Sigma^+ \longrightarrow \eta + \pi^+$$
$$5 \longrightarrow 4 + 1$$

$$\Sigma^0 \longrightarrow \Lambda^0 + \gamma \qquad \left( \gamma \longrightarrow \pi^+ + \pi^- \right)$$
$$6 \longrightarrow 4 + 2 \qquad \left( 2 \longrightarrow 1 + 1 \right)$$

$$\Xi^- \longrightarrow \Lambda^0 + \pi^-$$
$$5 \longrightarrow 4 + 1$$

$$\Xi^0 \longrightarrow \Lambda^0 + \pi^0$$
$$6 \longrightarrow 4 + 2$$

| | | | |
|---|---|---|---|
| $\gamma$ | gamma ray or virtual photon | $K$ | kaon |
| $\overline{\upsilon}_e$ | electron anti-neutrino | $\rho$ | proton |
| $e$ | electron | $\eta$ | neutron |
| $\upsilon_u$ | muon neutrino | $\Lambda$ | lambda |
| $\mu$ | muon | $\Sigma$ | sigma |
| $\pi$ | pion | $\Xi$ | xi |

It must be kept in mind that our vecton-formed particles exist as shells of force lines. Therefore, some cloud numbers might be able to sustain two or more states of layering. Hence, the positive kaon is basically a proton with fewer force line shells and a negative xi is a "fat" negative sigma.

Do all typical particle decays support this cloud theory? Well . . . yes and no. The exceptions I have found involve neutral particles whose existence in the formulas below may be due more to conservation speculation than to reality. Here are some examples: (the first line is the quantum decay version and the second line is vecton decay version).

$$K^+ \longrightarrow \mu^+ + \upsilon_u$$
$$K^+ \longrightarrow \mu^+ + \gamma$$

$$\Lambda^0 \longrightarrow \eta + \pi^0$$
$$\Lambda^0 \longrightarrow \eta + \text{neutrinos} \quad \text{or } \Sigma^0 \longrightarrow \eta + \pi^0$$
$$\text{or even } \Lambda^0 \longrightarrow \pi^0 + \gamma + \gamma$$

It is believed that all baryons in decay always yield a baryon. I cannot determine whether the neutron in the above lambda decay is there because it was actually observed or there because modern physicists think it should be. Hence, I would not rule out

$$\Lambda^0 \longrightarrow \pi^0 + \gamma + \gamma$$

even though baryon number is not conserved. Vecton theory does not require the conservation of baryon number or even its existence.

This magnet view of particles also explains the fixed masses of the various particles. If the particle is a composite of force lines, and if such force lines envelop the particle as shells in the same way the magnetic lines of flux surround a bar magnet, then it follows that some layers might be geometrically more stable than others. For example, each of the three clouds of a charged kaon would contain fewer force lines than each of the three clouds of a proton.

With this I have satisfied myself at least that a prima facie fit does exist for my view of particles. However, before I begin to really quantitatively study my cloud model of particles, I should first see if a similar first face argument can be made for the four attractive forces.

Therefore, let us now briefly look at the four attractive forces in nature, determining whether or not our vecton theory can offer any plausible generalizations which may explain them.

## IMPLICATION NUMBER TWO

## ALL OF THE ATTRACTIVE FORCES IN NATURE ARISE FROM NEGATIVE VECTORING AND THE EXCHANGE OF VECTONS.

### The Weak Force

We are taking the weak force first since we have already touched on it with our speculations on particle decay. The weak force is the most short-ranged force and the third weakest of the four known attractive forces in nature. Even so, it is some $10^{25}$ times as strong as the gravitational force. In the most general terms, it can be said that the weak force is operative in maintaining certain particle structures. For example, the muon decays into an electron through the weak force, as does a neutron into a proton.

In some ways the weak force is the least understood of the four forces. In modern quantum theory the weak force is believed to arise from the exchange of particles with mass called intermediate vector bosons. In quantum theory all of the four attractive forces arise from the exchange of particles. However, the actual descriptive mechanics of how particle exchange can effectuate attraction are not understood. Our theory cannot totally subscribe to the particle-exchange model of attraction. Of course, since particles are made of vectons, and vectons are exchanged to bring about attraction, the two concepts are not completely incompatible.

In discussing the decay of the muon, we suggested that the particle began as a fairly closed vecton system of force lines. The system then randomly lost vectons due to its own innate imperfections: imperfections associated with the inability of the outer lines of force to maintain themselves after a certain amount of vecton loss.

This means that to vecton theory, the weak force is the holding pattern forced on vectons in a line of force by negative vectoring. The decay of this pattern of alternating positive and negative vectons through vecton loss ultimately leads to the collapse of the entire line-up. This is the nature of pion and muon decay as well as the nature of neutrino formation. In the case of the decay of a neutron into a proton (beta decay) the same kind of line of force collapse takes place. However, in the case of the neutron the vecton loss is achieved a little differently.

It has long been a belief of quantum theorists that an electron is "created" on the surface of a neutron during a beta decay. Since the energy-mass of the departing electron and the mass of the remaining proton do not together equal the mass of the original neutron, an electron neutrino was postulated to account for the missing mass. Quantum theory is a little vague on why a neutron decays. Most explanations fall back on the law of probability. Explanations for the actual mechanics of mass-to-energy conversion are lacking.

The vecton cloud theory brings to beta decay a new interpretation for electron formation by the neutron. In our theory the neutron is initially "created" from a hydrogen atom by its interaction with another hydrogen atom, an accomplishment not quite as difficult in vecton theory as in classical theory due to our disbelief in unit charge.

If our galaxy began as a great cloud of hydrogen gas, and if electrons and protons are perfectly stable due to closure, then it follows that the forming of atoms more complex than hydrogen must conserve vecton cloud number. Would it not be possible for an electron, caught between two protons suddenly bound by the strong force, to be drawn against its proton partner? Now let us further suggest that this "unnatural" alliance can be preserved only as long as the proton-proton bond which originally established it is maintained. How the electron is held by the proton to form a neutron will become clearer when we discuss the strong force. However, we will state that the bond is related to the lines of force of the proton.

An analogy is needed at this point to explain our next concept. Usually, a log in a fireplace cannot burn by itself. Other logs must be placed around it so that some of the energy is "recycled" allowing the surface of the log to maintain oxidation temperature. It may be that two protons (one a neutron) are needed to establish a system of interaction capable of maintaining lines of force strong enough to hold an electron in parallel alignment as a fourth cloud for the neutron. This strength would be derived from the sharing of force lines. The neutron, upon leaving the nucleus would still have its lines of force operative, but without the sharing factor might randomly lose enough vectons to allow its lines of force to degenerate. The weaker lines of force would be unable to hold the captured electron and it would depart, freeing additional vectons to depart as an unobservable mass loss. Hence, the electron is held to the nucleon by the strong force and our weak force. A free neutron has only the weak force and eventually loses its grip on the electron.

Whether or not this particular generalization for beta decay and the nature of the weak force will be able to hold up under meaningful quantitative investigation is not the real question at this time. What is important is that vecton theory can at least address itself to the question of the cause of the weak force. Now, let us turn to the strong force to see if vecton theory can make any meaningful generalizations about it.

## The Strong Force

If a stable particle is indeed a kind of composite of circular force lines, then any two vecton clouds, which share force lines, would in effect become one particle. It would seem that such an intimate relationship could not be easily broken.

The strong force arises, according to this theory, from the sharing of force lines. This same sharing of force lines would also exist from nucleon to nucleon in a nucleus. In this theory, protons and neutrons are the same basic particle. If we think of a neutron as being an upside-down proton, we might conceptualize a tritium nucleus thusly:

### The Vecton View of a Simple Nucleus

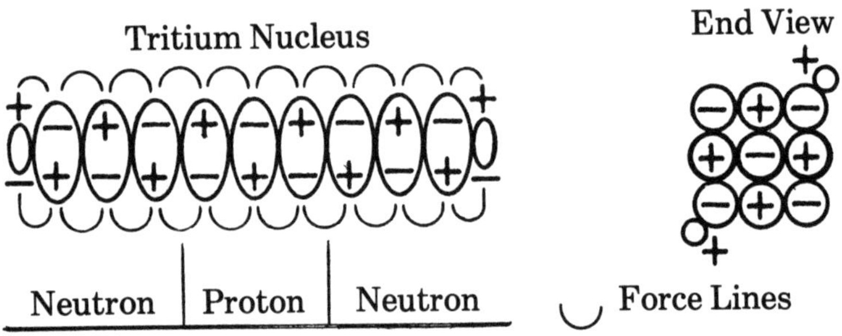


Figure 20. Vecton clouds in both particles and nuclei fit together like permanent magnets. This explains why protons don't repel each other in nuclei.

## Gravitation

On the shelf in my study sits a book on gravitation nearly 1300 pages thick. It's an impressive thing. Its arguments are very quantitative and have good fits with the popular theories. With careful comprehensive reading it will reveal most of the modern interpretations of the nature of gravity. Its only problem is that upon completing that work, you will still no more know the cause of gravitation than when you began. Such is the state of modern methodology.

Our vectons and vecton particles offer a very simple explanation for the cause of gravitation. Its one oddity is that it is easier to explain how a photon is gravitationally affected than how particles are. When we get to electromagnetic-wave propagation, we will explain how photons are formed. For the time being we will say that a photon may be envisioned as two clumps of vectons: one purely of positive vectons and one purely of negative vectons. In Figure 21 we portray a negative clump of 18 vectons in perfect equilibrium with no direction and a net speed of zero. In reality, the clump would contain perhaps billions of vectons and would only be in such directional equilibrium under the most uncommon circumstances. This clump is idealized for explanative simplicity.

**An Idealized Vecton Clump in Equilibrium**

Figure 21.

According to the logical implications of our postulates, if one negative vecton joined this idealized clump, the clump would slowly drift with the direction of the added vecton at a speed of 1/19th, the theoretical speed of light (Figure 22A). Conversely, if one of the vectons is removed from the clump through negative vectoring (Figure 22B), the clump would move in the direction of the lost vecton at a speed of 1/17th the speed of light (individual vecton speed).

### The Effects of Positive and Negative Vectoring on the Idealized Vecton Clump

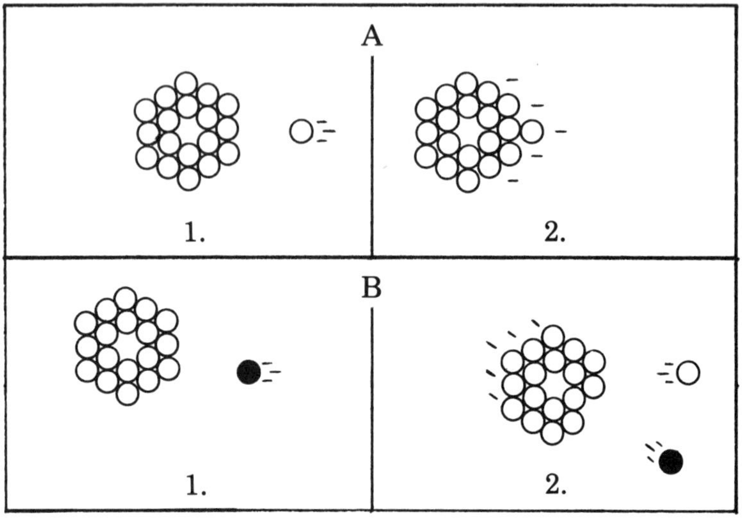


Figure 22. Positive vectoring (A) adds the influence of one vecton to the clump. Negative vectoring (B) subtracts the influence of one vecton from the clump. In a nonidealized clump, the negative vectoring vecton would probably dislodge more than only one vecton while positive vectoring must always offer only a unitary influence.

If all particles are comprised of vectons, and if these vectons exist in the particles through a probabilistic inability to escape a force line because of negative vectoring (the weak force), then it would seem that probability would also require a random loss of vectons from all particles. Therefore, if vectons are a reality, then all matter randomly looses individual vectons. The number of vectons lost would be directly proportional to the number of particles in a given mass, and the probability of finding a vecton from that particular mass would vary inversely as the square of the distance from it.

Now it would seem that our idealized cluster would not be moved either toward or away from a gravitating body. Since it likely would be struck by as many positive vectons as negative vectons from the direction of the gravitating body, its equilibrium would be maintained through averaging.

However, a real vecton clump would not be so idealized and its symmetry would be probabilistically determined rather than geometrically determined. Hence, the vectons in the clump interacting with the incoming vectons would not always be pointing exactly away from the vecton source. In a real photon the inside directions of the vectons would all be fairly parallel.

### The Gravitational Effect

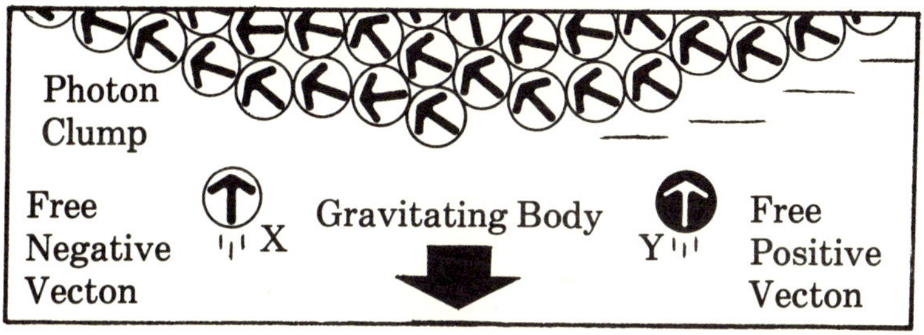


Figure 23. Incoming vectons (X and Y) from a gravitating body will interact with the vecton clump, X positively and Y negatively.

To an incoming like-charactered vecton this modest directional variety of the vectons in the clump would mean nothing since it would form a new vector sum with the entire body. But to an oppositely charactered vecton, which interacts vecton by vecton, the meaning would be significant. Remember, in negative vectoring a vecton always has the opposite direction and character of what it had before the interaction. This means the opposite incoming character could possibly rattle through the clump, transferred from vecton to vecton, like a bullet ricocheting in a steel drum. (Figure 24.)

**The Effect of Negative Vectoring on a Vecton Clump**

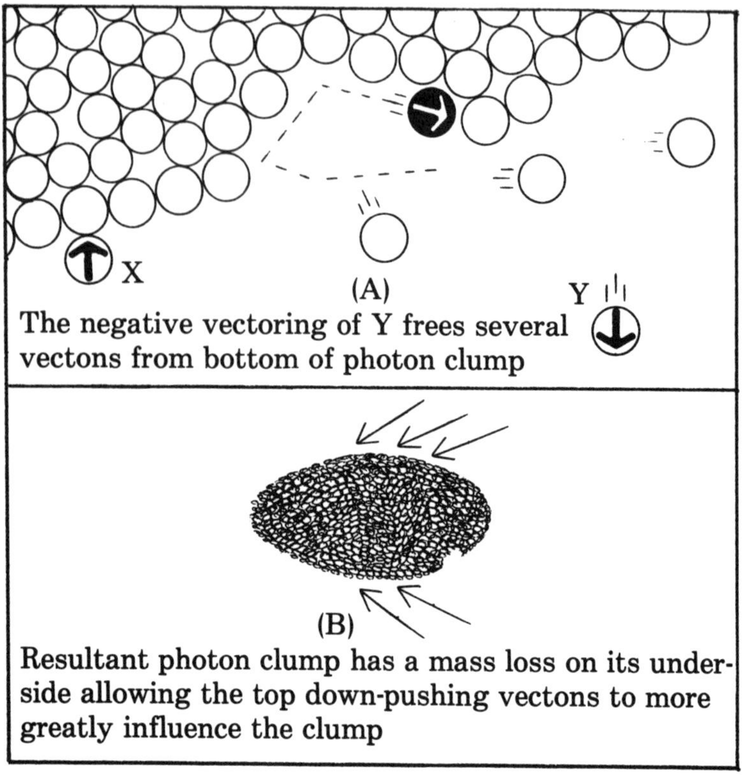


The negative vectoring of Y frees several vectons from bottom of photon clump

Resultant photon clump has a mass loss on its underside allowing the top down-pushing vectons to more greatly influence the clump

Figure 24.

Hence, like-charactered vectons may individually influence a vecton clump only to the degree of one vecton. However, oppositely-charactered vectons would have some probabilistic individual influence greater than one. What this influence might be can only be conjectured in this qualitative outline. However, it would have to be greater than one, and thus allow negative vectoring to more influence the clump than positive vectoring.

Gravitation for the photon, then, is caused by individual vectons "scratching away" their "underbelly." This model, therefore, cannot allow photons to have spin. It also means that gravitation causes a mass loss for the particles it acts upon and that all photons would have to be slowed when passing near a gravitating body as net vecton unidirectionalism is compromised.

If the original speed of the clump exists as a vector compromise among like-charactered vectons, and if these vectons have a common origin (discussed later), then the unidirectionality of the photon clump would be compromised by the influence of the like-charactered vectons it admits to its membership when passing near a gravitating body. Since our model requires that the speed of all photons and particles be a function of its vecton unidirectionalism, all photons passing near a gravitating body would be red-shifted.

It would seem that the influence of individual vectons on particles would be essentially the same as for photons. In our model vectons do not circle a particle but simply move back and forth having one character while in one direction and the other while in the opposite direction. Therefore, only character spins; the vectons themselves stay in place, held in the force line by the probability of being negatively vectored before being able to escape.

Our particle models all contain both positive and negative vectons. This distribution is exactly even in both neutral and charged particles. Charged particles have an even distribution of vectons in that charge arises purely from the direction of that character having the same direction at the core as the particle does as a whole. Hence, an electron at rest is a neutral electron, and if reversed, is a positron. Just as for a photon, the net speed and net direction of a particle arises from the net directional alignment of all of its vectons.

To an incoming vecton all of the rules apply to its encounter with the body of a particle as in the case of the photon clump, except for one. Each vecton entering a particle has a 50-50 chance of being negatively vectored.

This means that either character will strike an opposite character at first contact about half of the time. Hence, the number of "first touch" positive and negative instances of vectoring will be the same for both photons and particles.

However, the gravitational effect arises from the "scratch factor," the ability of the negatively vectored vecton to keep on being negatively vectored, or to at least pass on its disruptive opposite character. In the case of the like-charactered clump, one character of the two was always assured of a clear field of opposite vectons. In the case of a particle, that field for secondary instances of negative vectoring is cut in half because the vecton mixture is 50-50. Hence, our model implies that particles are gravitationally affected the same way photons are but only to half the extent.

It would seem also that particles must be slowed when passing near a gravitational field the same way photons are. For an orbiting body this slowing would be offset by a continuous acceleration brought on by the cumulative affects of negative vectoring. Our model requires all particles to continuously accelerate in a gravitational field because the "destruction" to the "underbelly" of every particle caused by each instance of negative vectoring is permanent and thus cumulative.

For those familiar with the tremendous penetrating power of neutrinos, the suggestion that individual vectons enjoy a similar penetrating power is understandable. A single vecton might possibly blow right through a light year of lead. Certainly our vecton explanation for the cause of gravitation is not compromised by the observed penetrating power of gravity.

The strong force according to this theory arises from the sharing of force lines which are themselves held together by the weak force. Both of these forces act over very short distances established by the particle's physical boundaries. By contrast the gravitational force is very long range.

## The Electromagnetic Force

The electromagnetic force, like the gravitational force, depends on negative vectoring and moves the particle through changes in direction. In some ways the electromagnetic force is the easiest of the four forces to understand.

Our view of a particle is as a permanent vecton bar magnet with positive vecton character flowing toward one end of the bar and negative vecton character flowing to the other end. Therefore, the character an approaching individual vecton or vecton clump (photon) must deal with depends on the direction it approaches the particle. For practical purposes we will ignore the outer force lines, treating the particle only in terms of its central core.

Only two possibilities exist for any one-charactered vecton clump or single vecton approaching a particle: either it will encounter like-charactered vectons or vectons of the opposite character. If it encounters vectons of the same character as itself, then it will simply form a vector sum with them. It will add its direction to the net direction of the particle, slowing it and changing its direction or accelerating it as the case may be. If the incoming vecton or clump encounters vectons of the opposite character, then it will move through the particle core, increasing the number of instances of negative vectoring. Since negative vectoring involves reverse directions, the particle will either be accelerated or decelerated by the directional influence of the incoming vectons. Figure 25 portrays a sequence illustrative of the way in which negative vectoring can bring about electromagnetic attraction.

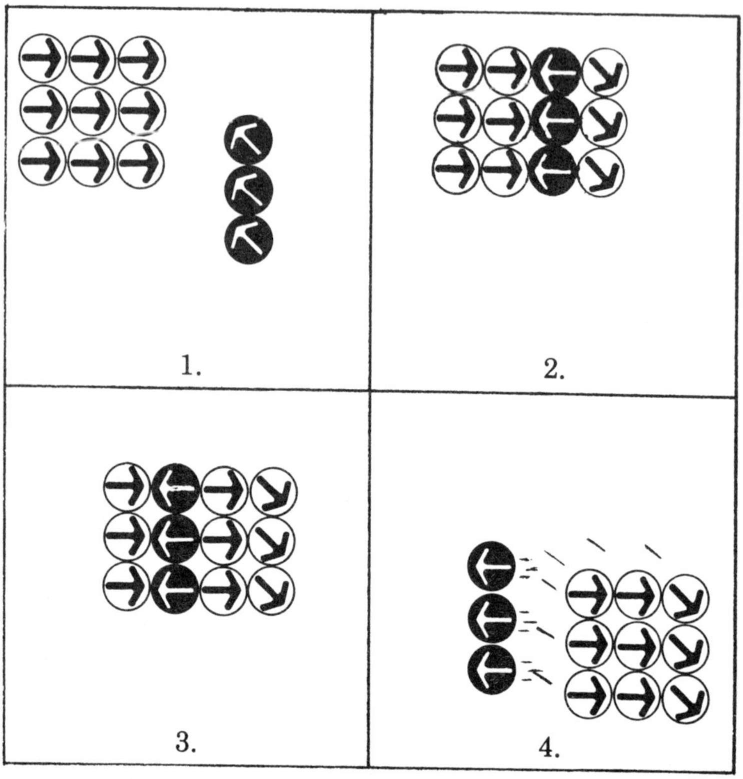


Figure 25. The electromagnetic attractive force arises when two oppositely charactered vecton groups exchange vectons and direction through negative vectoring.

Gravity is caused by individual vectons randomly emitted by matter. The electromagnetic force is caused by concentrations of like-charactered vectons, either as photon clumps or force lines. If our vecton model of the universe reflects reality, then it might be possible to determine the number of vectons in a photon clump simply by comparing the relative strength of the two forces. It may be that a single photon clump could contain as many as $10^{19}$ vectons!

This means that electromagnetic photon clumps would have great particle penetrating power. Whereas an individual vecton would only influence a few vectons at the particle's surface, a photon clump or force line of vectons could actually enter the body of the particle and thus be absorbed by it. Possibly the particle would absorb the photon bit by bit through negative vectoring in a way analogous to the way the atmosphere absorbs a meteor.

## Electromagnetic Wave Propagation

In classical physics the electron is bound to the proton through unit charge. In vecton theory unit charge is thought to be an illusion. Without unit charge we must now account for two things in atomic structure: (1) the force that holds the electron to the proton; and (2) the observations that one and only one electron is associated with each proton.

It is quite obvious that the electron must be bound to the proton by the electromagnetic force. Our model of vecton force lines makes this easy to see. The problem is that our model at first suggests that this bond would be polar, forcing the electron to the positive pole of the proton or forcing the forming of a four-cloud neutron. However, the reversed force lines of the middle upside-down quark or parton of the proton might serve to neutralize this potential unidirectionalism.

We must suggest, therefore, that there is a "zone of equilibrium" between the force line shells which allows the electron to travel perpendicularly to the axis of the proton. (Figure 26.) If each proton allows only one such zone, then it follows that each proton can hold only one free electron.

Since the force lines of the electron are charactered with respect to its direction, it follows that such a zone of equilibrium, if between two opposite force lines, might impart spin to the electron.

**The Hydrogen Atom**

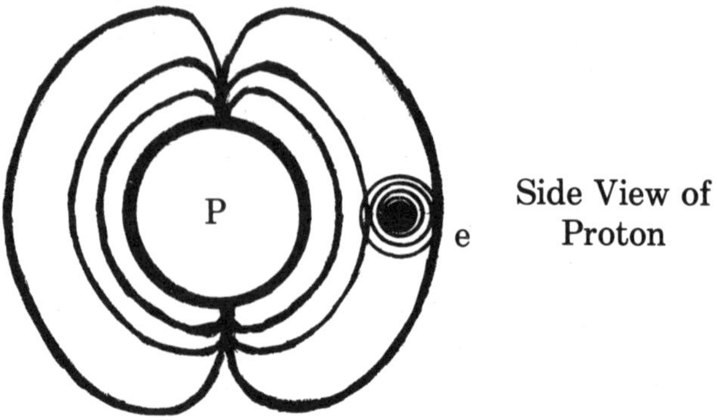


Figure 26. The electron orbits the proton's "equator" while being prevented from polar capture by an equilibrium in the force line zone of the proton created by the proton's one "upside-down" vecton cloud.

There may be an additional and more subtle reason for the one on one ratio. In this theory we came to the conclusion that a particle has a natural closure pattern requiring a least possible number of vectons. This number changes from cloud to cloud depending on factors such as force line sharing. Yet, each particle has its ideal size. Hence, all particles discard excess vectons. In the case of a hydrogen atom, the electron, according to our model, would be continuously absorbing vectons from the proton's force lines as it whirls around it. This means that one electron may monopolize the excess vectons of the proton and end the possibility that another electron could be sustained with an outer force line.

If vectons are energy, then they would be found everywhere in the same way energy is. Therefore, an energized proton would simply be a proton absorbing a quantity of vectons. If the proton does indeed hand its extra vectons to the electron through its force lines, what does the electron do with them?

As already noted, a particle cannot forever go on absorbing vectons. In our contemplations on the possible cause of particle decay, it became apparent that there must exist a most stable pattern of probabilistic vecton closure. This would seem to be that of a particle with an even distribution of vectons in its force lines. We concluded that excess vectons are emitted at either end of a particle's core because this would be the area of the particle where the likelihood of negative vectoring would be least likely to occur for any given vecton. In the case of an electron, a negative vecton has a chance of escaping out the front of the particle because there it would be surrounded by a majority of its own kind. A positive vecton would have a good chance of escaping from the other end.

Figure 27 portrays this theory's concept of an electron in orbit around a proton and shows the pattern of positive and negative vecton emissions.

**Electromagnetic Wave Propagation**

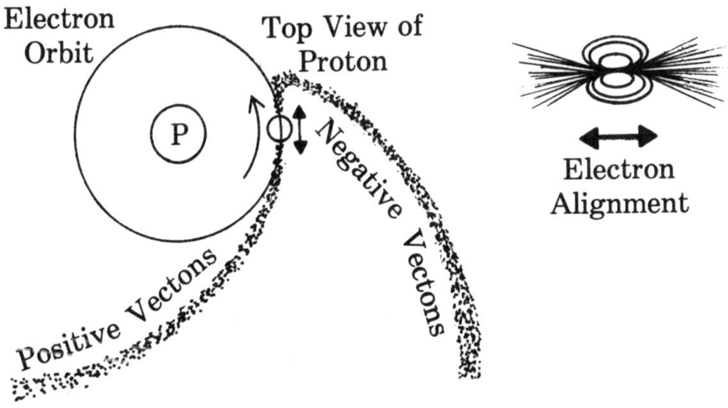


Figure 27. By continuously spewing negative vectons out its front end and positive vectons from its rear end, the electron sets up alternating sprays of positive and negative vectons which could coalesce into vecton clumps.

Thus, according to this theory, electromagnetic waves are just that, wave fronts with a frequency based on the electron's speed and distance from the proton.

If the electron is continuously absorbing vectons from the proton, then the electron's vecton spray must always be continuous. This means that, through the electromagnetic interaction, electrons in near proximity should affect each other, causing the electrons to align with each other through the electromagnetic force.

**The Magnetic Domain**

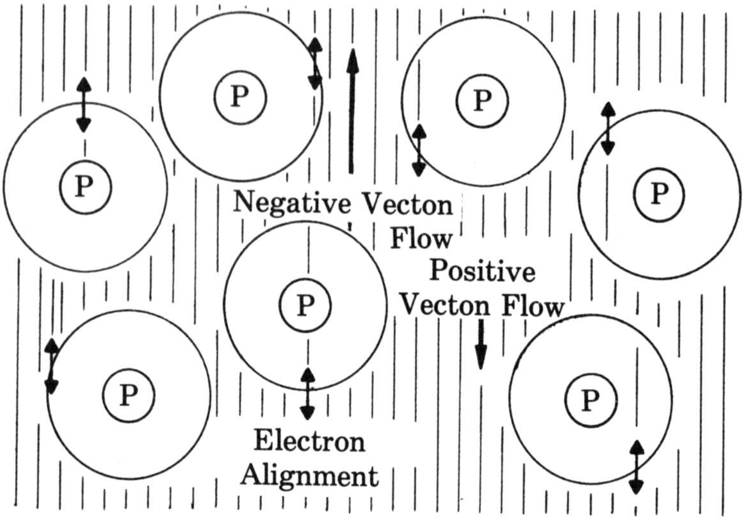


Figure 28. Electrons electromagnetically influenced by the vecton sprays of nearby electrons would align themselves accordingly. The resultant sympathetic alignment would cause all positive vectons to be sprayed in one direction and all negative vectons to be sprayed exactly oppositely.

In this situation each electron would, by aligning with the flow, also contribute to it. This we believe is the cause of permanent magnetism and the nature of a magnetic domain. Each permanent magnet then emits positive vectons at one end and negative vectons at the other. Further, it behaves as a giant particle, establishing force lines in much the same way particles do.

These force lines conserve the magnet's vectons at a level of natural equilibrium for the magnet's mass. Additional vectons are added to the mass by the impingement of matter (force applied). These "extra" vectons are then passed off from the magnet by vecton character at its charactered ends. Such is the nature of electrical and radio-wave generation. This is why a magnet must be put in motion before it can yield energy.

Vecton theory, thus, can approach and even unify the four attractive forces in nature. However, quantitative study should not be begun on this much alone. The vecton model of the photon needs to be studied for its fit with key experiments in physics. For example, does a vecton photon exhibit interference? The nature of waves must be dealt with. Let's now turn our attention to this matter.

## IMPLICATION NUMBER THREE
## ALL ELECTROMAGNETIC WAVES EXIST AS GEOMETRIC PATTERNS OF VECTONS.

### Vecton Theory Wave Definition

*A wave is a linear geometric positioning of discrete units whose pattern is maintained by a common time of origin, a common point or line of origin, and a common speed.*

### Planck's Constant

Our model of a hydrogen atom, as depicted in Figure 26, departs dramatically from the Bohr model, and probably, if vecton theory ever gets as far as step three, would undergo some structural change. With this model we have an electron evenly spraying vectons from both poles as it orbits the proton. Hence, the concentration of vectons will probably be the same at all points along the spray front. This is most likely because the electron would emit the same probable number of vectons for each orbit.

The logic for this last statement goes this way: if the emitted vectons are the electromagnetically accepted vectons from the proton's force lines, and the speed of the electron remains unchanged, then the number of vectons sprayed in each orbit is the measure of the proton's influence. If the proton is energized, its force line will swell with vectons, drawing the electron ever closer. The electron speed would stay constant, but its frequency of orbit would increase proportionally to its vecton emissions. Hence, each orbit would emit the same number of vectons. This means that an observer at a distance from the nucleus would always receive the same number of vectons per measured wave front regardless of how energized the proton. Only the frequency of vecton arrival in his instrument would fluctuate with the energy of the proton-electron system.

Does this mean that Planck's Constant measures the influence of a single vecton or sequence of single vectons? Quite likely a single vecton, or even thousands of vectons are not detectable. Also quite likely, the vecton spray does not remain homogeneous but breaks up into probable units or clumps of vectons through positive vectoring. Probability would impose a sameness on these wave front clumps which would serve as the unit of influence of Planck's Constant. The constant clumps would form at the poles of the electron.

If our even vecton spray front probabilistically breaks up into units, then white light might be envisioned as we depict it in Figure 29.

**White Light**

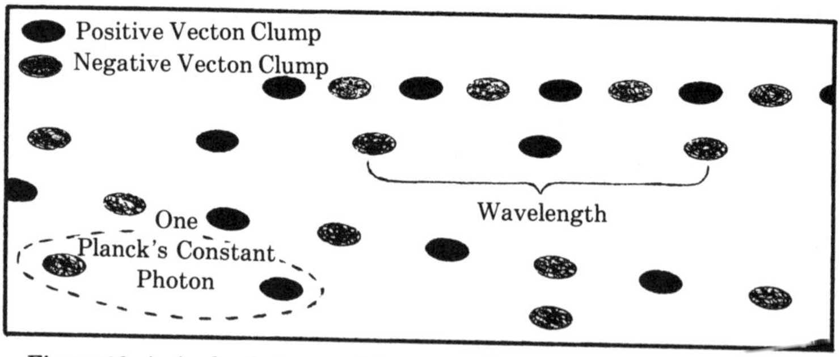


Figure 29. A single photon would consist of two vecton clumps, one positive and one negative. Wavelength would be defined by the distance between two like-charactered clumps in a series.

## Diffraction and Refraction

Earlier we suggested that our photon clumps might enter particle force lines in a manner analogous to the way in which a meteor enters the earth's atmosphere. However, it must be pointed out that vecton loss through negative vectoring would be offset by vecton gain through positive vectoring. Therefore, we might expect a photon not directly hitting a nucleon to retain its approximate size even though the vectons it has upon exiting the mass might not all be identical with the ones it entered with.

We established that vecton speed is a function of the net parallelism of all of the vectons in a photon clump as is net direction. Therefore, all photons upon entering matter must be both slowed and directionally altered as their vecton membership changes. If the mass is solid or deep, the vecton photons have a great chance of being absorbed by a nucleon, "chewed up", and resprayed by the electron of the receiving atom. However, if the matter offers hallways created by crystalline structures or has short depth, many photons will escape nucleon absorption and be able to exit the matter in their newly acquired speed and direction. This model, therefore, suggests that diffraction and refraction arise from essentially the same cause.

In refraction a directional equilibrium is imposed on the photon clump as it passes through the matter involved, which almost always exhibits some geometric atomic ordering. Therefore, a photon upon entering the matter will not be directionally altered while passing through translucent matter but will be greatly influenced in terms of direction when the mass it is interacting with does not equally surround it. Hence, the photon would be most bent upon entering the matter and upon exiting it.

## Dispersion

In our vecton model of reality, dispersion is brought about by the same factors that cause long wavelengths to have more penetrating power than short ones. According to this theory, light travels as a sequence of alternating positive and negative vecton clumps. We have no mechanism for allowing one vecton clump to be influenced by another. How then can matter affect different wavelengths in different ways?

If we return to our meteor analogy, we see that all photon clumps are to some degree torn apart by negative vectoring when they enter matter. If they enter a nucleon they are probably completely taken apart down to the last vecton. This is quite a feat considering that one photon clump might contain more than $10^{19}$ vectons.

Considering for a moment those clumps not absorbed by a nucleus, we see that they must leave behind themselves a trail of vectons of their own character as they interact with particle force lines and the vecton background that surrounds all atoms. Since the clump immediately following it will always be of the opposite character, its instances of negative vectoring will increase in direct proportion to the distance it follows the clump in front. If too close, or the matter too dense or deep, the short wavelength photons might completely interfere with each other, slowing so much that they are captured by the ambient matter.

In the case of a prism, the various wavelengths associated with white light are separated from each other by this factor. The longer wavelengths are not as self-interfering as the shorter ones and so are bent and slowed primarily by the atoms in the glass. The shorter wavelengths are bent more because they not only have to contend with the glass but with the disintegrating clump in front.

### Fraunhofer Lines

Assuming we are dealing with a uniform gas, we see that if the spacing (wavelength) of a given photon alignment corresponds with a common electron orbit of the gas, when one vecton clump is struck by an electron, every clump in the lineup will be similarly "clunked" as they pass by. We may assume all photon lineups will lose some members in this way, but those whose wavelength match the orbit will lose all of their members. Hence, certain wavelengths simply can't make it through certain gases!

### Interference

It is quite apparent that our vecton clumps are not unlike Isaac Newton's Corpuscles. Yet, they have one very important characteristic that neither Newton's corpuscles nor Quantum theory's photon has: they exhibit positive vectoring. Our photons can form vector sums. With this feature vecton theory can deal with both the observed wave nature of light and its particle nature.

For a moment, let's liken monochromatic light to Pickett's charge at the battle of Gettysburg in 1863. General Pickett's men marched forward in lines like a picket fence. They had a common pace and common line of origin. To the Union troops a mile and a half away, they probably looked like a series of grey waves. Yet, their wave nature was an illusion created by a patterned march of individual men.

The idea that light travels as discrete units of energy was first proposed by Isaac Newton. This idea was declared implausible by an experiment conducted by Thomas Young in 1801. In this experiment, Young passed visible light through two closely-spaced holes. The light set alternating bands of dark and light on a screen behind the apertures establishing "proof" that light has wave properties.

Most of quantum mechanics, uncertainty included, is strongly influenced by the assumption that photons in collision would deflect away from each other. However, if they form vector sums, as our vecton photons do, then interference bands would be a natural product of corpuscular light.

In the Young interference experiment, two factors must be kept in mind. Firstly, the two windows must always have an aperture which permits diffraction. This alone accounts for the secondary wave set up by the experiment. Secondly, wave reinforcement is nothing more than the forming of vector sums.

**The Young Interference Experiment**

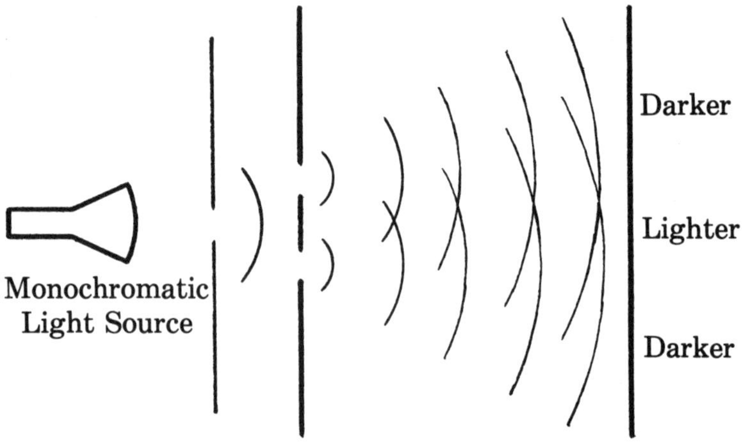


Figure 30. In vecton theory, the interference bands are explained by the positive and negative vectoring of photon clumps. Like-charactered clumps reinforce each other while the collision between clumps of opposite character are self destructive, scattering free vectons.

Physicists have been so married to the Huygen's concept of waves that when it was found that electrons also exhibited interference, they opted to believe that electrons exist as probability waves in spite of overwhelming evidence of the particle nature of electrons. What should have occurred instead was a re-examination of the assumptions associated with wave theory.

Our vecton particles, of course, would explain the photo-electric effect as well. Quite possibly the vector summation of two or more photon clumps might be needed to account for the influence (energy) necessary to eject an electron.

### The Hubble Red Shift

At the time of this writing it is popularly believed we live in an ever-expanding universe. Further, the expansion seems uniform in all directions. The basis for this belief rests upon an observed red shift in the spectrum of light arriving from far off galaxies. In 1929 the astronomer Edwin Hubble showed there was a correlation between these red shifts and the distances of their galactic sources: the more the red shift, the farther away the galaxy. Astronomers believe these red shifts are Doppler shifts. This implies the galaxies are flying apart as the result of a central cosmic explosion (the "big bang") which formed the universe about fifteen billion years ago. However, a universal radiation back-drop, first observed in 1964, is also uniform in every direction and shows no bias toward a "hotter" center of an exploding universe.

Has modern astronomy returned us to the Ptolomaic notion that we are at the center of the universe? Is it not plausible that the observed directional uniformities of the red shift and the radiation better imply a steady state universe than an expanding one? Could the Hubble red shift be, for instance, a "gravitation shift" instead of a Doppler shift?

In discussing gravitation we observed that all matter randomly emits free vectons and that vecton photons would have their unidirectionality, and hence speed, compromised by gravitational influence. The degree of that compromise would have to be directly proportional to the square of the distance the photon passes from the gravitating body.

Our vectons, once emitted from matter, travel indefinitely until absorbed. Therefore, all of space is full of free vectons. Certainly our vecton photons would probabilistically gather and discharge vectons through positive and negative vectoring with these free vectons. Such exchanges will affect the parallelism of the vectons in a photon clump and it will be slowed in proportion to the distance it must cover. Hence, the farther a photon travels through space, the more it is red-shifted.

Another peril for photons in space is collision with other photons. Such random and probable collisions could probabilistically eliminate clumps from photon sequences, further increasing the red shift.

**Black Holes**

If vecton clouds form from concentrations of vectons; if these concentrations go on to form simple and complex particles as hydrogen atoms; if these hydrogen atoms gravitationally coalesce to form more complex atoms, stars, and planets, as a galaxy; then what will a galaxy become? A galaxy will eventually become a huge concentration of vectons: a black hole.

A black hole is the ultimate particle and a natural stage in the evolution of all galaxies. Every galaxy will eventually form a black hole at its center. In this black hole, held together by negative vectoring, no electromagnetic waves can be formed. Only free vectons can escape its surface. Since free vectons are the cause of gravitation, the black hole becomes an awesome gravitating body.

The black hole is not a hole at all but a huge mass spewing its vectons to every part of the universe. Eventually these free vectons will join others to form simple vecton clouds and eventually new galaxies with new stars, new planets, and new life.

If vecton theory is correct, then we are part of an amazingly simple and ordered universe: its smallest unit of expression being the vecton and its largest the galaxy or galaxy group. Suppose the Hubble red shift is not a Doppler shift . . . then the radiation backdrop that surrounds us in every direction may not be from a big-bang at all but the light remnants of distant galaxies beyond which are more galaxies forever and ever.

# Verifying the Genesis Model

### (Steps Two and Three)

The genesis model we have just presented has all of the qualities a good model should have. We called this model Vecton Theory, taking license with the word theory since no quantitative evidence was offered. Yet, look at what it can explain.

In its metaphysical development, Vecton Theory left no unanswered internal "whys." Why direction exists, why straight lines exist, why absolute speed exists, why conservation exists in physics and chemistry, and why vector sums form are only a few of the answered whys. In contrast, modern theory cannot say why unit charge exists, why a photon always travels at the speed of light in a vacuum, or even why mass and energy are interchangeable.

In the generalizations we made with regard to its implications, Vecton Theory was able to account for the observation that mass and energy are essentially the same thing and the observation that the speed of light is independent of its source; it unified the four attractive forces and brought logical perspective to the wave-particle duality of light. Finally, it was able to offer an alternative to the Doppler explanation for the Hubble red shift, freeing us from some of the implausibilities associated with the big bang theory. Not bad for only two tiny entities!

With this we now initiate step two where we begin to put our prima facie fits into quantitative perspective.

Phase two of step one only establishes that the idea of a vecton universe is worth some investigative effort. Further, it only establishes that suggestion for me personally. New ideas, especially qualitative ones, are like unruly children: only their parents can really love them.

The parent of any reality model must realize that with step one he has nothing that can be publicly presented until his quantitative homework is done. Therefore, he has two options. One is to quantitatively examine his ideas, taking the implied mathematics of his model and working them into fits with the observations of certain key experiments. The other option is to concentrate on one novel prediction of the model.

Working with the first option in the case of the sample model, I might take the fact that my cloud numbers are always conserved in most typical modes of particle decay and quantitatively develop the implied particle descriptions. From this I might be able to determine the number of force line shells the various particles have, their relative size, and perhaps even venture a guess as to the number of vectons found in an electron at rest.

In a sense, the first option would be a quantifying of the speculations of step one phase two. The problem with this is that step one is too inductive and too qualitative to be trusted to reward such a massive investigative effort. Inductive methods are only economical when their per effort investments are low.

Therefore, I suggest that this first option be held in abeyance as a step three and that step two concern itself with one novel prediction of the model which can be quantitatively tested. Therefore, the product of step two would be a rigorous paper outlining an experiment which would yield a prediction conforming to that of the model but not to any other existing theory.

The choice of experiment would, of course, vary from reality model to reality model. In addition to making unique predictions it should be comparatively inexpensive. The correct choice of experiment for step two is very important.

In the case of the sample model, two possible candidates for step two are evident. One relates to the prediction that light slows as it forms nonparallel vector sums and the other turns on the prediction that weak decays yield free vectons.

If vectons exist, as the model suggests they do, then light undergoing interference would travel at a slower speed after interference than before. If it is possible to set up a Young interference experiment with apparatus capable of measuring the speed of monochromatic light from source to dual window slits, and then from windows to screen, the model's prediction could be tested.

The problem with this test is that the angle of vector summation would be too small to yield dramatic speed differences. Further, the speed of light is not easily measured in conjunction with the Young apparatus. This experiment might not be able to yield conclusive results.

The experiment relating to weak force decay is not only more meaningful but much more exciting. If the weak force arises from a probability of inability for vecton escape in force lines, then weak force decays should yield free vectons. If in turn free vectons interact with particles to cause the gravitational force, then the following statement can be made:

> All weak force decays induce a gravitational influence directly proportional to the mass loss involved and inversely proportional to the square of the distance from the point of decay.

An experiment based on this prediction would have three basic elements: a good source of energy release involving weak decays, a screening device to filter out high energy particles and photons, and a gravity sensing device.

Nuclear reactor cores are a hotbed of weak decays. However, the unidirectionalism associated with particle force lines means that the vectons would leave the particle in closely associated groupings with a character bias (neutrinos, if you prefer), rather than in randomly scattered free and evenly mixed vectons. The problem would be to get close enough to overcome the inverse square dilution to obtain measureable results while not exposing the experimenters to dangerous radiation. Even so, the neutrino vecton groupings may not yield a uniform gravitational influence.

The problem presented by this experiment is not unlike the problems presented to those seeking to discover neutrinos in the 1950's. Extraneous particles must be filtered from the energy source so that the particles comprising the gravity sensing device are not "pushed" by photons and other energy release debris. This would require thick lead walls between the reactor core and the sensing device.

Finally, there is the matter of the sensing device itself. Fortunately, several gravity sensing devices have already been built for the, to date, fruitless search for gravitons. Gravitons are the carrier of the gravitational force in quantum theory.

The primary task anyone has with step two is a correct interpretation of the model's selected prediction. The experiment must be carefully and quantitatively outlined. If the experiment fails, it must do so because of error in the model's prediction, not error in calculation.

If it so turns out that the unique prediction of the model is verified by rigorous experiment, then the model, cosmology and all, is ready for public presentation and the area by area scrutiny already mentioned.

A successful step two will not prove the model but will make its ideas acceptable for the investment of time and money necessary to carry out step three.

Because the uniqueness of this methodology is found almost entirely in step one, steps two and three have only been touched upon. Essentially, they represent nothing more than conventional scientific method. Step two would be a paper outlining the physical theory of the model in quantitative terms while defining a meaningful experiment which would verify a unique implication of the theory. Step three would involve quantitative review by the scientific community of all the predictions of the model if the results of the experiment of step two are positive.

# The Ontology

Ontology as used here is the study of being and awareness. Historically its primary interests have tended to center around the nature of mind and soul. With man's growing understanding of animal intelligence ontological speculations have, in modern times, been extended to the entire spectrum of biology. Basically, the ontologist attempts to answer the question: What is awareness and mind? Here, we will also treat the basic question of epistemology (how can a being have knowing?) as an aspect of our ontology.

For the materialist, awareness is a direct consequence of the electrical interplay between atoms, electrons, and molecules. For him both mind and soul are the manifestations of a hard-wired neurological system which handles memory and new information in a manner not incomparable to the way modern computers handle similar functions. Therefore his ontology would flow from his matterology. Because the physics and chemistry of neurology are fairly well understood and not concerned with the mysteries of sub-particle physics, a materialistic ontological model can be developed independently from a genesis model. That is, since all viable genesis models must imply a chemistry identical with the chemistry we know, that section of all reality models will be the same. Therefore, any ontology consequential to the laws of chemistry will automatically be consequential to any viable genesis model. The nature of the laws of chemistry is just about the only aspect of reality on which there is universal human agreement.

For the dualist, however, his cosmology is vital to his ontology because he cannot substantiate the existence of a supernatural soul or spirit with "natural" physics and chemistry. That is, as soon as he argues that awareness results from the same stuff that atoms and molecules are made of he becomes a monist. The following diagrams outline the basic difference between a monistic reality model and a dualistic one.

The Monistic View of
Consequentialistic Progression

Cosmology → Matterology → Ontology → Conductology

The Dualistic View of
Consequentialistic Progression

Cosmology → Ontology → Conductology
Cosmology → Matterology

It could be argued that all matter is really a manifestation of spiritual substance. The mass-energy relationship of special relativity popularized this view among those who believe that energy and spirit are one. However, matter and spirit are only meaningful terms when dualistically perceived. If energy is the ultimate stuff of the universe, then it is meaningless to argue whether it is spirit or material, even if it comprises the body of a Creator God. This much can be said: A spiritualist may be either a monist or a dualist, but all materialists are monists.

Because our sample reality model is a materialistic one, we should have some overall outline of our matterology before commencing our ontology. So far we have been treating consequentialism as both a scientific method and a philosophical method. From this point on we will be speaking purely philosophically, excepting for the following brief matterological review.

Our genesis model gave us a cosmology of absolute substance characterized by only four things: infinite empty space, positive vectons, negative vectons, and the vecton properties. This means that every structure in our matterology can have no substance unique to itself, but must draw its identity from whatever relative positioning of vectons it sustains. We found that energy existed in three vecton interaction forms and matter in only one. Energy exists as free vectons, as like charactered vecton clumps (photons), and as groups of noninteracting mixed vectons traveling in a common direction (neutrinos). The fundamental form of matter is that of a set of closed vecton interactions which set up circular force lines. These vecton clouds or bags are the leptons and quarks of particle physics.

This view allows us to categorize particles as being either simple (possessing one cloud) or complex (having more than one cloud). Accordingly, a hydrogen atom arises from the combination of one simple particle and one complex particle. The same would be true for a neutron. This model thus allows the symmetry that the number of positive vecton clouds in the universe exactly equals the number of negative vecton clouds in the universe. By contrast, unit charge theory has the disequilibrium that there is about 1,836 times as much positive mass in the universe as there is negative mass.

Our vecton matterology does not contain any matter of its own at all, but rather an array of increasingly complex patterns of vecton clouds which are themselves formed by the relative positioning of vectons. Therefore it is appropriate to outline our matterology from the standpoint of basic pattern formulation and degree of complexity.

It is known that immense clouds of hydrogen exist in the universe. Quite likely these clouds are the progenitors of the galaxies. Since a hydrogen atom is the simplest combination of one stable complex particle and one stable simple particle, we may say that atoms form a new level of interaction. Molecules, the next level of pattern, would then form in consequence to a gravitational collapse of the hydrogen into eddies from which arise the solar systems. This, then, is the qualitative progression of pattern complexity as viewed from the vecton perspective.

Positive and negative vectons
give rise to
Simple particles which in turn yield complex particles
together they give rise to
Atoms which in turn yield molecules
together they give rise to
Landscape

In this case we will use the term landscape in its broadest cosmological sense. We will let it cover all of those things covered in geology as well as in classical astronomy. Comets, suns, planets, volcanoes, oceans, cliffs, and minerals, are all expressions of atomic and molecular patterns. That is, they are chemically simple. However, when we view them as an interacting group, then a new level of complexity is introduced. As an interacting system, the components of landscape form an environment. For example: an orbiting moon tugs at an ocean of water which then washes against a cliff of calcium carbonate to change the chemical composition of the ocean water. Thus, just as the interaction between atoms gives rise to molecules, so does landscape interaction give rise to environment. In this sense, landscape and environment together constitute all of what we consider to be the material structures of our universe.

Just as each new category in vecton evolution contains all of the properties that precede it, so does each plateau in the matterology contain all of the achievements of reality that precede it. The environment arises from landscape interplays, the landscape arises from the existence of atoms and molecules. Atoms arise from complex particles and simple particles, which in turn arise from vecton interaction. Vecton interactions arise from vecton relativity and so on to the void of volume. Such is the nature of consequentialism.

Our environment, therefore, contains every aspect of reality simpler than itself. It holds photons, possibly ninety-two elements, free electrons, and an array of molecular combinations limited in diversity only by the fundamental law of probability. Hence, some environments will contain, probabilistically, some very complex molecules.

Molecular bonding, according to vecton theory, does not arise from the sharing of electrons, but from the sharing of force lines. Every particle force line has a certain capacity to hold the vectons of other force lines. This capability forms a kind of probabilistic trap which serves as a pseudo-appetite. For example, it would be inappropriate to say that a single oxygen atom needs a hydrogen molecule. Yet, when exposed to the hydrogen, the oxygen behaves as if need truly existed.

In the case of very complex molecules existing in an enriched environment the same illusion of need exists. A certain benzene molecule, short one carbon atom, will lie in wait with its empty force line serving as the web to catch the carbon fly. Further, the positioning of the force line will be such that only carbon will be "eaten" and other atoms and molecules will likely not be.

To a chemist this molecular appetite turns on the concepts of valence and charge. To the bio-chemist it lies at the very heart of a complex molecule's ability to duplicate itself. To understand this, view for a moment all molecules as pluses and minuses. We will make the rule that all pluses "need" minuses and vice-versa. We will further suggest that the best bonding pattern exists when each plus or minus is shared with two of its opposites, but that other bonds are possible. (Figure 31.)

## Molecular Replication

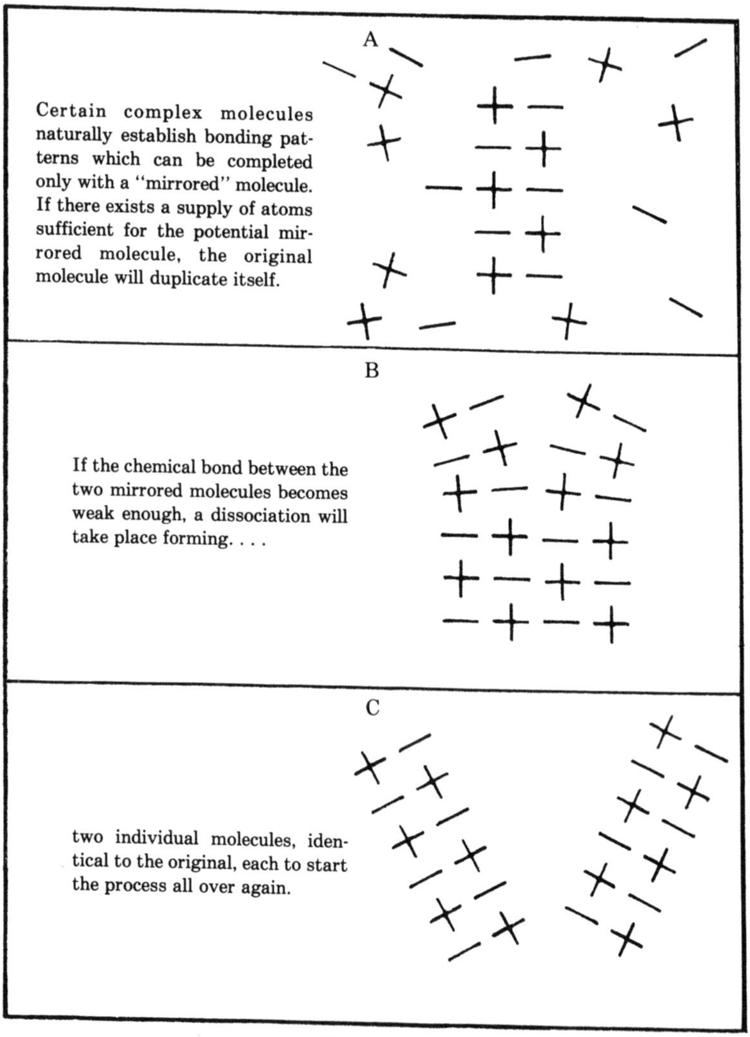


Figure 31.

DNA and RNA, the molecules of life replication, are much more complex that our simple plus and minus analogy. However, they operate according to just about the same kind of principle. Quite likely, also, the earliest forms of replicating molecules were much simpler than the complex life molecules we now have through evolution.

The most fundamental difference between nonliving and living things is in the ability to self-duplicate. With this distinction then we will draw the line between those things we will consider matterologically and ontologically. We will call the ability to replicate, life principle.

Life principle creates three basic needs for the complex molecules so qualifying. The first is the need to "eat," to acquire additional atoms and molecules for the task of structure building and replication. Since even simple life forms are selective in their diet needs, those life systems which are able to contact many molecules and reject those they do not need will better survive than those lacking this ability. Therefore, excretion may be regarded as the second need created by life principle. Finally, there must exist the need to reproduce. At the simplest levels this need might exist simply as a chemical "need" for molecular dissociation when a certain molecular chain has been completed.

Not every replication will form a perfect duplication. Chain B in Figure 32 formed a molecular hook which tends to trap the plus and minus "food." Therefore, the pattern of Chain B will perhaps survive better than that of Chain A.

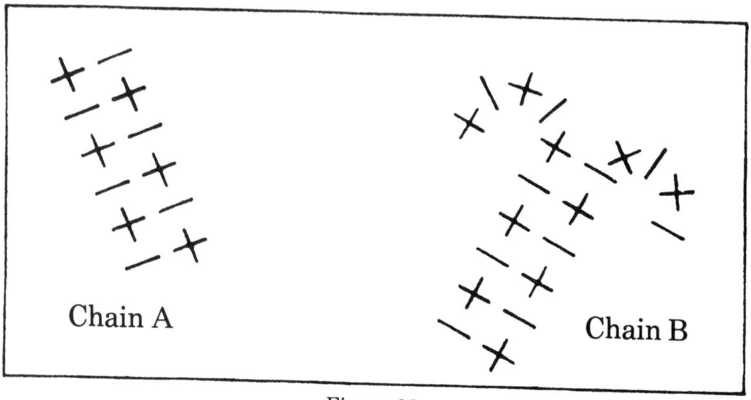


Figure 32.

Even the simplest viruses and bacteria best thrive on food of complex molecular structure. Therefore, they need compounds which are already organic in nature. Plants have developed the ability to use less complex compounds. Those living things lacking the talents of plants to survive on the basic chemicals at hand needed the ability to sense complex molecules. At first, this might have been nothing more than a tendency to grow in the direction of a source of complex nitrogen molecules. With the advantage of locomotion, the being could utilize other clues given off by organic compounds, such as heat or vibration. We will call this ability to sense and identify a source of desired compounds sensory perception. In the case of higher life forms this would involve the use of all five senses and the servicing of needs beyond those of life principle.

Associated with the achievement of sensory perception is the development of a nervous system. Most plants have not evolved to the level of sensory perception because their ability to utilize water and other simple chemicals have not made it essential. Yet, the existence of carnivorous flora does suggest that some plants have some sensory development.

Even at the level of mammals sensory perception is a very physical thing. All sight is stimulated by the impinging of photons on highly specialized tissue. Sounds are transmitted by energy patterns involving the momentum of mass. Smell and taste rely on cells with certain chemical affinities, and touch involves the transfer of energy through pressure. Any sight, sound, taste, smell, or touch can be duplicated simply by duplicating the neurological energy patterns it creates.

Memory, therefore, is essentially the ability of a nervous system to duplicate a given pattern of electrical impulses. In vertebrates this is achieved through a highly developed system of neurons and their attendant synapses, axons, and dendrites. Each synapse serves as a kind of off-on switch, not unlike those used to store information in a computer. If the information in an incoming electrical impulse is able to trigger a certain neuron route, then that route will serve as a pseudo-pattern of the original impulse pattern. Such is the nature of memory in the most cursory possible terms.

With memory, living beings are able to evolve from simply sensing their environment to reaching an awareness of some of its portents. Memory constitutes a new level of achievement for reality. We will call this new plateau the level of environmental awareness.

As intelligence is selected-in to a species and an awareness develops about the existence of the material world, memory will begin to provide personal insight as well. Self-identity is simply a composite of memories and sensory experiences. Strangely, self-awareness does not come from within, but from without. This is why humans tend to see themselves as they believe others see them. We do not decide for ourselves whether we are ugly or good-looking, intelligent or stupid, entertaining or dull, or privileged or impoverished. It is natural for each of us to take our cue on these things from others.

As intelligence further increases, pattern recognition develops. At this level, life is able to make predictions about things never experienced on the basis of pattern. Logic, as pattern recognition, then is a new achievement of reality. With both personal awareness and logic, individuals will develop a sense of perspective.

Since our matterology is defined by varying degrees of pattern complexity, we are treating our ontology from the pattern perspective also. That is, we are not concerned with specific form, but with classes of complexity. A being capable of logical thought requires more interaction of matter and energy than a being which has only sensory perception.

To this point our ontology consists of five plateaus of complexity, or achievement levels. They are life principle, sensory perception, environmental awareness, self-awareness, and logic. Each of these levels is characterized by a basic set of needs.

For life principle the need is to serve the appetites of bodily function and reproduction, as well as to excrete. For sensory perception the need is for pleasure, not pleasure fully in the hedonistic sense, but perhaps more in the sense of freedom from discomfort. Interestingly, this ontology does not give us a desire for self-survival as a direct need. That is, we don't take care of ourselves to survive, we take care of ourselves because it hurts when we don't. Many humans, especially young ones, will happily risk their lives in reckless adventures whose only purpose is to fulfill such needs as sexual appetite or recognition service.

At the level of environmental awareness the basic need is for security or freedom from fear. One of the first things life found out when it learned there was a world out there was that it contained things which cause pain.

For the level of self-awareness the most basic need is for recognition. If our view of ourselves depends upon what others think of us, then it is very important that others think well of us. One aspect of the need for recognition is the need to dominate. Man, killer whales, dogs, and chimpanzees all exhibit unnecessary aggressive behavior. Both dogs and man will kill for an insult. It may be that man owes his logical mind to an intense competition for dominance within his own ranks which put selective pressures on intelligence. It is one thing to survive by outsmarting a less intelligent prey, and quite another to survive by outsmarting your biological equal. Homo Sapiens could be the sole survivor of a half-million years of intraprimate warfare and aggression. The last of the Neanderthals died (were killed off?) about thirty thousand years ago and, even now, the gorilla and orangutan are threatened with extinction by human poachers.

Self-awareness, therefore, creates intraspecies competition which in turn creates selective pressures favoring intelligence and pattern recognition. With pattern recognition comes logic and the need for order. What is music, but the ordering of sound patterns, or geometic design, but the ordering of light patterns?

Out of self-awareness and the need for order comes the highest achievement of reality as far as we humans know it, the plateau of moral behavior. When man's mind so expanded that he realized on an individual basis that others have feelings and awareness just as he did, then he developed empathy. At this stage he wants things to be right for others, as well as for himself. At the level of altruism he cares, and he seeks moral order.

The achievement of moral order is a comparatively new thing for man. Indeed, the genetic characteristics it requires may not even yet be universally held in man. A look at religious thought, which seems to have always epitomized the highest ideals of man may be instructive. The neolithic religions had very localized gods whose primary functions were to be the objects of intense praise and to successfully direct one's spear against one's enemy. These gods operated very much at the level of recognition need and personal awareness.

As the need for order grew, religion was called upon to explain things. Therefore, the religions of the early post-neolithic period attempted to provide explanations for such things as river flooding, lightning, rainbows, lunar periods, and even the origin of man and his sense of personal awareness. The political systems were highly theocratic and what laws of moral behavior that did exist were usually not extended to strangers. For instance, Moses, after providing the ten commandments, saw nothing morally wrong with ordering the slaughter of perhaps thousands of helpless non-Hebrew women and children (Numbers 31).

Then, at some point less than three thousand years ago, there emerged in Asia the idea that love and compassion were to be extended to all men. The teachings of both Confucius and Buddha were highly moralistic. These teachings spread westward, adding their ideas to the moral philosophies that had developed in the Middle East and southern Europe. Finally, there emerged the two great religions of Christianity and Islam whose followers worship a universal God of love for all men. Today, altruistic concern is being extended by some, not only to fellow humans, but to other mammals as well.

What I have attempted to do here is to explore, in superficial fashion, a new and very different way of classifying living things. Instead of looking at life in terms of physical structure, I have evaluated it qualitatively, seeking to define those things all life have in common at various levels of evolution. In matter, the different levels of complexity are easily defined because they occur in unit jumps. The difference between an atom and a molecule is a quantitatively distinct one. However, the evolution of living things forms an unbroken continuum which forces all qualitative evaluations to be somewhat arbitrary. Yet, improved awareness capabilities are directly linked to increased sophistication in the physical structures of nervous systems so that qualitative evaluation is not without quantitative foundation.

Indeed, it might be argued that such ontological interpretations of life might be the only meaningful way to compare the various stages of evolution. Such a classification as the one presented in this model might have universal applicability while physical classifications might be inoperative for creatures from other solar systems. Even on this planet awareness levels might better serve to classify life. Human muscle tissue, gram for gram, is inferior to that of many animals while our ability to grasp objects is shared by the most humble of tree monkeys. Is a bush really inferior to a sponge just because it is a plant and the sponge is an animal? No other class of multi-celled living things has the natural genetic adaptive capabilities of the class insecta. Are mammals therefore their inferiors?

Of course, if you decide to develop an ontology of your own you might want to approach it from a different perspective. However your model of awareness, life, and being is formulated, it should have a logical consequentialistic linkage which develops a hierarchy from the simple to the complex. You may wish to evaluate life in terms of longevity, spiritual awareness, species adaptability, physical size, DNA or cell complexity, or even your own interpretation of awareness levels. As long as it provides you with a view of being that makes sense to you, then it is sufficient.

# The Conductology

Conductology is a term coined by this author to cover all of the "ics" of philosophy. Economics, politics, ethics, and aesthetics all to some degree deal with the relationships men have with each other, with other living things, and with their environment.

You may or may not wish to deal with these things in a reality model of your own. Because I wish to cover all possible aspects of a reality model, I am at least mentioning them. More than likely you will prefer to treat these subjects from purely a personal perspective, attempting to define your attitudes about these subjects in light of the implications of your reality model.

For instance, in the case of the sample model, the conductology would represent nothing more than the proper response to the needs generated by the six levels of awareness. Here are some typical conductological speculations I have made from the sample ontology.

If the nature of man is as our ontology says it is, then the conduct of man is fairly predictable. Man will always act in accordance with his needs. However, different needs can motivate identical behavior. For instance, an individual's considerate behavior in a particular case might be motivated by a fear of recognition loss, a logical understanding that it might be the best means to achieve a more fundamental need, or an honest altruistic need to be concerned. Therefore, political, economic, and legal systems might be structured to encourage moral behavior in a society by rewarding more fundamental needs, especially recognition need.

To what degree should government be altruistic? A nation which restricts its economic productivity out of concern for the environment and its military strength out of humanitarian concern could more easily be subjugated by another nation lacking such values.

The risks of self-destruction brought on by altruism exist even at the individual level. Perhaps this is why man has not yet been able to evolve beyond this level. The man who jumps into the river to save the drowning child may himself drown while the unsympathetic onlookers will survive to pass on their genetic code. In modern societies we see many individuals restricting the number of children they produce out of a sense of parental responsibility to not have more children than they can afford and, sometimes, out of a concern for world overpopulation. Others, having no such concerns, have all the children that happen to come their way. The problem is furthered as those with fewer children, out of empathy for the impoverished children of irresponsible parents, tax themselves to support children not carrying their own genetic code. This further reduces their ability to responsibly have children while encouraging the child bearing of those parents not supporting their own children. With what system of ethics does one approach these kinds of problems?

From this perspective, it would seem that any conductology predicated upon our ontology must not only be oriented to need service, but to the preservation of upward genetic selection pressures; that is, to preserve whatever genetic need for moral order might exist in a society. Because this book serves only to explain the nature of consequentialistic thinking, I will not explore here what system of economics, law, ethics, and even aesthetics, might achieve these and other goals implied from the ontology. I will only point out that, from the vantage point of this model, modern concerns over moral government, religious values, world peace, good citizenship, economic fairness, and general morality, are expressions of the highest possible achievement reality possesses.

A proper conductology is not essential to a good reality model. Yet it is important that an individual's political, economic, and ethical viewpoints are in tune with his views of the nature of being and the nature of matter. He should have some logical basis for his attitudes about the things around him. Should whales be treated as an economical source of oil and fertilizer, or as a fellow form of very intelligent life? Man cannot properly deal with the world around himself unless he has some way of evaluating its various aspects.

# The Conclusion

It is the goal of the consequentialist to provide for himself, and others who follow his thinking, a logical and unified view of reality. A good model should have no unanswered whys or internal contradictions. The vecton theory offered here is such a model. As to whether it accurately reflects the nature of reality is a question which must remain unanswered until its quantitative implications are more fully developed and put to rigorous experiment. The important thing is that it is able to offer prima facie explanations for those popular beliefs and explanations which are contrary to its views.

The veracity of this particular model of reality is not really at issue in this book. What is at issue is its ability to show that consequentialistic methodology and perspective can at least address, in some meaningful way, the major mysteries of our universe. There is but one reality, and only when we come to understand the full scope of that reality will we come to know whom we really are.

Though I do believe that intelligent life is to be found elsewhere in our galaxy, I am inclined to think the evidence for "flying saucers" is too circumstantial and gossamer to be accepted without serious reservations. Yet, I am struck by a story by a Dr. Daniel Fry, who claimed in 1950 that he was taken aboard a flying saucer and given an explanation of its workings.

He is quoted as being told: "The force which accelerates the vehicle is identical in nature to a gravitational field. It not only acts upon every atom of the vehicle itself but also acts equally upon every atom of mass which is within it, including the pilot and passengers."

Fry's visitors from another solar system went on to say that our laws of physics are much too complex. "Your greatest need," Fry quotes the visitors, "is to discover the utter simplicity of the basic laws or facts of nature . . . then you will be able to produce effects which now seem to you impossible. Go back down the limb on which you are trapped to the point where it joins the main trunk and then start up again."

This story came to my attention just prior to the printing of this book and I hastily insert it here because of its fit, not only with the approach of our methodology, but also with our sample reality model as well. Let me explain:

If vecton theory is correct, then the well-defined destructive interference of photons will yield free vectons. Free vectons are the cause of gravitation in vecton theory. This means that any vehicle could be accelerated, steered, and braked gravitationally with nothing more than monochromatic light beams and a specialized interferometer! In any case, the methodology of consequentialism certainly constitutes a climbing back down the limbs of Quantum Theory and General Relativity Theory. Who can say what new and simpler views of the laws of nature we will find when we start up again?

Here, then, is the vecton model of reality in its entirety.

# THE VECTON MODEL OF REALITY

THE COSMOLOGY or GENESIS MODEL
operative laws
and
The Law of Zero
The Law of Infinity

the universe begins as a
VOID of
UNIVERSAL NOTHINGNESS
which expresses
LAW as
the law of INFINITY and the law of ZERO
which are mutually expressed as
SPACE
which is
UNCONFINED VOLUME and CONFINED VOLUME
which are in turn mutually expressed as
DIRECTION
which is
POSITIVE DIRECTION and NEGATIVE DIRECTION
which in turn are mutually expressed as
DIMENSION
which is
INFINITE DISTANCE and FINITE DISTANCE X SPEED-TIME
which are in turn mutually expressed as
MOTION
(THE VECTON)
which is
POSITIVE CHARACTER (POSITIVE VECTONS) and NEGATIVE CHARACTER (NEGATIVE VECTONS)
which in turn are mutually expressed as
RELATIVITY
which is
POSITIVE VELOCITY and NEGATIVE VELOCITY
which are in turn mutually expressed as
INTERACTION
which is
POSITIVE VECTORING and NEGATIVE VECTORING

THE MATTEROLOGY
operative law
The Law of Probability

which are in turn mutually expressed as

PHOTONS (ENERGY) and FORCE LINES

(positive vectoring) (negative vectoring)

which are in turn mutually expressed as

SIMPLE PARTICLES which form COMPLEX PARTICLES

which are in turn mutually expressed as

ATOMS which form MOLECULES

which are in turn mutually expressed as

LANDSCAPES which form ENVIRONMENTS

from conducive environments arises

THE ONTOLOGY
operative law
The Law of Natural Selection

LIFE PRINCIPLE and the need for appetite service

from which arises

SENSORY PERCEPTION and the need for painlessness

from which arises

ENVIRONMENTAL AWARENESS

and the need for security

from which arises

SELF-AWARENESS and the need for recognition

from which arises

LOGIC and the need for order

from which arises

ALTRUISM and the need for moral order

# Glossary

**Absolute Speed**—A speed which is relative only to itself and to which other speeds must be relative. Vecton speed. The speed of time in Vecton Theory. The speed of light.

**Absolutism**—In philosophy: the view that reality rests upon invariable laws, structures, and principles.

**Achievement**—In Consequentialism: the classifying term for a defined level or stratum of complexity in a consequentialistic reality continuum.

**Aesthetics**—In the vecton reality model: man's interpretation of the value of various aspects of his environment. Classically: philosophical studies on the nature of beauty.

**Astrophysics**—The branch of physics studying the fundamental causes of astronomical phenomena.

**Beta Decay**—The radioactive decay of a neutron in which an electron is liberated.

**Black Holes**—Popularly: a hypothetical mass whose escape velocity exceeds the speed of light. A collapsed star. In Vecton Theory: any cloud of vecton concentrations with sufficient vecton density to prohibit sustained particle force lines. The decay product of a galaxy from which only free vectons are emitted.

**Boson**—A classification term in physics for particles whose spin is zero or an integral number. Those particles which, according to quantum theory, bring about the attractive forces.

**Clump**—In Vecton Theory: a concentration of like-charactered vectons held together by positive vectoring. Half a photon.

**Conductology***—In Consequentialism: the study of conduct as implied from a given ontological perspective.

**Consequentialism***—A philosophical method of investigating reality with the use of deterministic reality continuums. The area of physical philosophy which attempts to discover metaphysical truths through the logical implications of observed truths.

**Cosmogony**—The study of primary cause and origin. A Genesis model.

**Cosmology**—A theory about the fundamental principles of the universe.

**Determinism**—The philosophy that events and even willful selection are always resultant from antecedent causes. It differs from consequentialism only in that classically it has not tended to concern itself with the structural nature of matter and energy, nor the concept of reality models.

**Dimension**—In Vecton Theory: the mutual expression of two equal and opposite directions as an infinite distance and a finite distance having infinite time.

**Dispersion**—The separation of light into a wave-length continuum. The "break down" of white light into colors.

**Diston***—In Vecton Theory: the quantum unit of finite space or volume. Absolute space or distance. The least possible measure of volume greater than zero. The amount of distance-volume expressed by one vecton.

**Dualism**—The philosophy that reality exists in two irreducible and separate parts, usually spirit and matter.

**Duration**—In Vecton Theory: the measure of all the motion that takes place within any finite set of vectons while a single vecton travels the length of one diston.

**Empiricism**—The belief that reality can only be understood through experience.

**Ethics**—The branch of philosophy dealing with the study of moral principles.

**Event**—In Vecton Theory: any specified pattern formed by any specified set of vectons. Any specified relative positioning of vectons. Also: a point of time, a pattern, a relative positioning, or a structure.

**Fatalism**—A deterministic philosophy which holds that all events are predestined by fixed laws governing the existing state of reality.

**Finite**—Popularly: that which can be measured. In Vecton Theory: that which is least possible or any real number multiple of that which is least possible.

**Force Lines**—In Vecton Theory: circular systems of vecton interaction involving enough vectons of both character to insure the probability that any given vecton will be returned toward the center of the system by negative vectoring before escaping. The weak force.

**Frame of Reference**—In Vecton Theory: any finite set of vectons placed within the context of two discrete events.

**Fraunhofer Lines**—Dark bands in a diffraction spectrum created by the energy absorption patterns of various atoms.

**General Relativity**—Albert Einstein's Theory of Gravitation advanced in 1916.

**Genesis Model***—In Consequentialism: A cosmogony.

**Hadron**—In Vecton Theory: a particle consisting of two or more vecton clouds. In Quantum Theory: a particle comprised of quarks. Mesons and baryons.

**Infinite**—Without limit, unbounded. In Vecton Theory: the natural measure of a direction's point. Unlimited size.

**Infinity**—The quality of being infinite. In Vecton Theory: having universal existence. Unlimited count.

**Interaction**—In Vecton Theory: the achievement of reality governing the rules of vecton collisions.

**Interference**—In physics: alternating bands of energy concentrations and energy nulls created by wave merger.

**Lepton**—The light particle group which includes the muon, the electron, and the neutrinos.

**Logic**—An ontological capability to recognize pattern. The twenty-first achievement of reality in Vecton Theory.

**Materialism**—The belief that reality can be fully explained with only matter and energy.

**Matterology***—In Consequentialism: the study of matter and energy.

**Meson**—Any strongly interacting particle between the mass of a muon and the mass of a proton. In Vecton Theory: positive and negative pions are classed with leptons since they decay through the weak force.

**Metaphysics**—The study of those things beyond direct observation. Typically: cosmological and ontological speculations.

**Methodology**—A particular method of investigation.

**Monism**—The belief that reality is predicated upon only one irreducible property.

**Motion**—In Vecton Theory: (1) the product of a unit of finite distance expressing itself over a unit of infinite distance so that the two distances are made equal as required by the law of zero. Because distance is the sum of two equal and opposite directions, there are two kinds of motion, each characterized by one absolute speed and one of two opposite characters. Speed plus character. (2) The vectons.

**Mutual Expression***—In Vecton Theory: the essence of physical coexistence or summation between two equal and opposite properties.

**Negative Vecton***—Half a finite dimension. A unit of negative motion expressing one direction, one of two opposite characters, one absolute speed, perfect rigidity, and irreducible volume.

**Negative Vectoring***—In Vecton Theory: a vector subtraction exemplified by the letter V. The interaction between two vectons of opposite character, resulting in their mutual exchange of character and a directional reverse for each.

**Neutrino**—In modern physics: one of six massless leptons which serve as the expressions of energy not imparted to the tau, muon, or electron at the time of their formation in particle decays. In Vecton Theory: a noninteracting group of vectons of one character, traveling in close parallel association from a common origin.

**Ontology**—The study of being and awareness. A particular theory on the nature of being.

**Parton**—In modern physics: a theoretical physical sub-structure of the proton, perhaps not synonymous with quarks; i.e., a proton contains three quarks and perhaps forty partons in theory.

**Pattern**—In Vecton Theory: any finite expression of relative distances.

**Photon**—In Quantum Theory: a unit of radiant energy. In Vecton Theory: two like-charactered opposite clumps of vectons with each clump held by positive vectoring.

**Physical Philosophy***—An area of study which attempts to approach the questions of theoretical physics with methodologies which are both qualitative and quantitative.

**Point of Time**—In Vecton Theory: see Event.

**Positive Vectoring***—In Vecton Theory: a vector summation exemplified by the letter Y. The interaction between two or more vectons of like-character which forces the individual vecton directions to be compromised as a resultant direction and the individual vecton speeds to be conserved as a resultant speed and a binding force.

**Property**—In Vecton Theory: any universal expression, value, entity, characteristic, or essence capable of influencing any other universal expression, value, entity, characteristic, or essence.

**Qualitative**—Expressing essence as opposed to measure. In Vecton Matterology: expressing pattern.

**Quantitative**—Expressing measure as opposed to essence. In Vecton Theory: a reference to those properties of reality which owe their existence to the absolute consequential aspects of a void rather than to any vecton pattern; i.e., infinite empty space, the positive vectons, the negative vectons, positive vectoring, and negative vectoring.

**Quantum Theory**—In modern physics: the belief and the attendant beliefs that electromagnetic energy travels as finite quanta which express both wave and particle characteristics.

**Quark**—In modern physics: a mathematical sub-particle construct which may allude to hadron behavioral modes only or hadron structures, or both.

**Rationalism**—The view that truth can be understood with only the tools of reason and logic.

**Reality**—All that is. The universe. Physical existence and being.

**Reality Continuum, Reality Model**—A logical and sequential ordering of both observed and theorized aspects of reality.

**Red Shift**—An increase in electromagnetic wave length, sometimes associated with Fraunhofer lines and sometimes with frequency change, typically believed to be caused by either the Doppler effect or gravitation.

**Refraction**—A change in direction for electromagnetic waves, associated with a change in medium for those waves.

**Relative Positioning***—In Vecton Theory: see Pattern and Event.

**Relativity**—In modern physics: Albert Einstein's theory (1905) that (1) light is comprised of photons which travel at an absolute speed; (2) the laws of physics are invariant in all inertial systems; (3) mass and energy have equivalence. An extension of this theory in 1916 to explain gravitation using tensorial geometrics. In Vecton Theory: the relationship at any given point in time between sets of vectons as determined by the relative positioning of the vectons involved, their directions, and their speeds. The determinator of interaction.

**Scientific Method**—A method for finding truth involving: (1) problem recognition, (2) hypothesis formulation, (3) prediction formulation, and (4) observation by way of experiment to verify the prediction.

**Speed Time***—In Vecton Theory: the measure of the obligation of a finite unit of distance to express itself over an infinite distance.

**Speed**—In Vecton Theory: the measure of absolute vecton motion disregarding character. Vector values predicated upon this measure.

**Strong Force**—In modern physics: the fundamental binding force for quarks and nucleons. In Vecton Theory: the holding force that arises between vecton clouds through the sharing of force lines.

**Sufficient Condition**—A set of properties or circumstances that assures the existence of a consequential set of properties or circumstances.

**Tautology**—In Consequentialism: A closed set of statements, laws, and/or properties whose members exist sequentially to one or more truly fundamental statement, law, and/or property also a member of the set and which is relative only to itself.

**Time**—In Vecton Theory: observed changes in the relative positioning of vectons ultimately caused by absolute individual vecton speed. Speed-time.

**Uncertainty Principle**—The tenet of Quantum Theory that a particle's position at any given time is always indeterminable because of its motion. (Vecton Theory does not take issue with this view of fundamental matter for the observer perspective. However, this theorist does feel that use of the uncertainty principle as an explanation for particle decay is not valid.)

**Velocity**—In Vecton Theory: speed as a function of a vector relationship. Relative speed.

**Weak Force**—In modern physics: the fundamental binding force for mass in certain instances. Beta decays, pion and muon decays, and neutrino formation all involve weak force decays. In Vecton Theory: the holding pattern for individual vectons in a force line brought about by directional vecton paralleling and the statistical probability that an incidence of negative vectoring will take place before the vecton has a chance to escape the force line.

**Zone of equilibrium***—In Vecton Theory: a theorized zone circumferencing the proton created by a balance between the spin direction of the proton's upside-down cloud and its two upright clouds which allows an electron to orbit the proton without polar attachment.

*Denotes words coined by the author.

# Index